Safety With Machinery

Safety With Machinery

John Ridley and Dick Pearce

OXFORD AMSTERDAM BOSTON LONDON NEW YORK PARIS
SAN DIEGO SAN FRANCISCO SINGAPORE SYDNEY TOKYO

Butterworth-Heinemann
An imprint of Elsevier Science
Linacre House, Jordan Hill, Oxford OX2 8DP
225 Wildwood Avenue, Woburn, MA 01801-2041

First published 2002

Copyright © 2002, John Ridley and Dick Pearce. All rights reserved

The right of John Ridley and Dick Pearce to be identified as the authors of this work has been asserted in accordance with the Copyright, Designs and Patents Act 1988

All rights reserved. No part of this publication may be reproduced in any material form (including photocopying or storing in any medium by electronic means and whether or not transiently or incidentally to some other use of this publication) without the written permission of the copyright holder except in accordance with the provisions of the Copyright, Designs and Patents Act 1988 or under the terms of a licence issued by the Copyright Licensing Agency Ltd, 90 Tottenham Court Road, London, England W1P 0LP. Applications for the copyright holder's written permission to reproduce any part of this publication should be addressed to the publishers

British Library Cataloguing in Publication Data
A catalogue record for this book is available from the British Library

Library of Congress Cataloguing in Publication Data
A catalogue record for this book is available from the Library of Congress

ISBN 0 7506 4830 9

For information on all Butterworth-Heinemann publications visit our website at www.bh.com

Composition by Genesis Typesetting, Laser Quay, Rochester, Kent
Printed and bound in Great Britain

Contents

Preface ix

Part I Safeguarding philosophy and strategy 1

1 Safeguarding of work equipment 3
  1.1 Introduction 3
  1.2 Design considerations 4
  1.3 'Cradle to the grave' concept 4
  1.4 Designer's responsibility 5
  1.5 Safeguarding principles 6
  1.6 Guarding strategy 7
  1.7 Standard-making bodies 8
  1.8 European approach to machine guarding 8
  1.9 Unit vs holistic approach to safeguarding 9

2 Factors affecting the selection and effectiveness of machine safeguards 10
  2.1 Introduction 10
  2.2 Basic factors 10
  2.3 Quality assurance 13
  2.4 Reliability 14
  2.5 Integrity 16
  2.6 Validation 17
  2.7 Summary 17

3 Typical hazards of machinery 18
  3.1 Identification 18
  3.2 Agents of hazards 18
  3.3 Hazards from parts of machinery and work equipment 20

4 Risk assessment, risk reduction and selection of safeguards 35
  4.1 Introduction 35
  4.2 What is a risk assessment 35
  4.3 Risk reduction strategy 36
  4.4 Determining a Safety Integrity Level for machinery hazards 40
  4.5 Selecting a safeguarding system 49
  4.6 Summary 53

Part II Guarding techniques — 55

5 Mechanical guarding — 56
- 5.1 Introduction — 56
- 5.2 Guard selection — 56
- 5.3 Guard types — 57
- 5.4 Other factors to consider — 68
- 5.5 Other techniques — 69

6 Interlocking safeguards — 71
- 6.1 Introduction — 71
- 6.2 Interlocking devices — 71
- 6.3 Guard locking — 75
- 6.4 Interlocking systems — 76
- 6.5 Levels of risk — 76
- 6.6 Interlocking media — 77
- 6.7 Two hand controls — 79
- 6.8 Hold-to-run controls — 80
- 6.9 Limited movement control — 80
- 6.10 Person sensing devices — 80
- 6.11 Lasers — 85
- 6.12 Pressure sensitive mats — 86
- 6.13 Pressure sensitive edges and wires — 88
- 6.14 Grab wires — 88
- 6.15 Emergency stop switches — 89
- 6.16 Telescopic trip switches — 90
- 6.17 Proximity switches — 91
- 6.18 Key exchange systems — 91
- 6.19 Key interlock switches — 93
- 6.20 Delayed start — 93
- 6.21 Other interlocking devices — 94

7 Ergonomic aspects of machinery safeguarding — 95
- 7.1 Introduction — 95
- 7.2 Physiology — 95
- 7.3 Controls — 96
- 7.4 Machine layout — 98
- 7.5 Colours — 98
- 7.6 Lighting — 99
- 7.7 Noise — 100
- 7.8 Vibrations — 100
- 7.9 Rate of working — 100
- 7.10 Temperature and humidity — 101
- 7.11 Ventilation — 101
- 7.12 Repetitive actions — 101
- 7.13 Warnings — 101
- 7.14 Vision — 101
- 7.15 Radiations — 102
- 7.16 Indicators and instruments — 102
- 7.17 Coda — 104

Contents vii

| Part III | Safeguarding systems | **105** |

8 Mechanical safety arrangements — 107
   8.1 Introduction — 107
   8.2 Guards — 107
   8.3 Distance fencing — 113
   8.4 Safety gaps — 115
   8.5 Feed and take-off stations — 117
   8.6 Work-holding devices — 118
   8.7 Counterweights — 118
   8.8 Safety catches — 118
   8.9 Braking systems — 119
   8.10 Clutches — 121
   8.11 Summary — 124

9 Electrical safety circuits — 125
   9.1 Introduction — 125
   9.2 Effect on safety — 126
   9.3 Basic safety requirements — 126
   9.4 Selection of interlocking switches — 127
   9.5 Switching contact requirements — 128
   9.6 Factors influencing the selection of interlocks — 129
   9.7 Circuit fault protection — 129
   9.8 Safety control circuits — 133

10 Hydraulic safety circuits — 153
   10.1 Introduction — 153
   10.2 Hydraulic systems for safety circuits — 153
   10.3 Hydraulic safety circuits — 154

11 Pneumatic safety circuits — 162
   11.1 Introduction — 162
   11.2 Pneumatic installations — 162
   11.3 Pneumatic safety circuits — 164
   11.4 Summary — 169

| Part IV | Other safety-related arrangements | **175** |

12 Safety in the use of lifting equipment — 177
   12.1 Introduction — 177
   12.2 Common safety features of lifting equipment — 177
   12.3 Additional features for particular lifting equipment — 178
   12.4 Lifting accessories — 182
   12.5 Circumstances requiring special precautions — 182
   12.6 Precautions when handling lifting equipment — 183

13 Safety with pressure systems — 184
   13.1 Introduction — 184
   13.2 Compressed air — 185
   13.3 Hydraulic installations — 190
   13.4 Steam — 193

14	Safe working with equipment	201
	14.1 Introduction	201
	14.2 Systems of work	201
	14.3 Protection from electric shock	202
	14.4 Locking off	203
	14.5 Ergonomics	205
	14.6 Anthropometrics	206
	14.7 Openings in guards	206
	14.8 Operating instructions and manuals	206
	14.9 Labels on equipment	207
	14.10 Supervision	207
	14.11 Use of jigs and fixtures	207
	14.12 Safety clothing	208
	14.13 Stored energy	208
	14.14 Signs and signals	208
15	Plant layout and the working environment	210
	15.1 Introduction	210
	15.2 Space	210
	15.3 Buildings	211
	15.4 Services	211
	15.5 Ventilation	211
	15.6 Lighting	212
	15.7 Temperature	212
	15.8 Machine layout	213
	15.9 Noise	213
	15.10 Vibrations	214
	15.11 Materials handling	214
	15.12 Maintenance	214
	15.13 Waste	214
	15.14 Access	215
	15.15 Lubrication	215
	15.16 Dust and fumes	215
	15.17 Floors and foundations	216
	15.18 Hygiene	216
	15.19 Notices and signs	216

Part V	**Appendices**	**217**
1.	Published standards	219
2.	Glossary of terms	227
3.	Abbreviations	232
4.	Smooth shaft pick-up	234
5.	Pipeline colour codes	235
6.	Permit-to-work	236
7.	Protection of enclosures	238

Index 241

Preface

Increasingly today the engineer, whether designer, producer or maintenance, is faced with more and more complex machinery in a society that is demanding higher and higher levels of protection for its members who have to operate machines. As a result there has been a proliferation of laws, standards and practices aimed at ensuring that an adequate level of protection is provided.

The designer, producer and works engineer is put in the invidious position of having to choose the most suitable and effective means to meet that level of protection commensurate with the demands of the process in which the equipment is to be employed.

This book sets out to simplify and clarify how those obligations can be met while remaining sane and economically viable. It explains the principles involved in the protection techniques and the methods of application of safety equipment and gives the engineer sufficient basic information to enable a rational and reasonable selection from the range of equipment offered to be made in the confidence that this decision will ensure that the degree of protection demanded by laws and society is achieved.

Essentially the book is something of a user's guide in as much as it starts with the basic guarding techniques required to protect against a single simple hazard and develops to encompass complex guarding problems associated with modern technically complex machines. For the more complex and technically advanced protection systems specialist advice may be required that can only be provided by the manufacturers of the protective equipment concerned. Much of the design and quality of the safety equipment is covered by European and international standards that are listed in an appendix.

The subject has been approached from a very basic level and developed to encompass the latest 'state-of-the-art' technologies in machinery guarding. Some of the illustrations used are of old machines with guards that have been in use for many years. We make no apologies for this, but only comment that a well designed robust guard should remain effective for the lifespan of a machine to which it is fitted. Experience has shown that the simplest guards are often the best, can be very effective and meet both the operator's and operational needs, as well as legislative requirements.

x   Preface

In preparing this book we have been conscious of the bewildering range of safety switches, equipment and circuits available for the engineer to choose from, so for clarity and to emphasize the safety features of a particular item or circuit we have included block diagrams to indicate the principles of operation of that equipment or circuit. These block diagrams highlight the safety features that the equipment must possess and, hopefully, they will assist the designer or engineer in identifying the essential features (or lack of them) in the equipment he/she is offered.

We have been fortunate in being able to draw on the expertise and facilities of a number of specialists and would like particularly to acknowledge the help we have received from Zurich Risk Services in the preparation of the chapter on steam generating equipment, from Michael Dray of Mackey Bowley International Ltd on hydraulic safety devices and Warren Ibbotson of EJA Ltd for diagrams of electrical and electronic safety devices. Other manufacturers of safety equipment have been generous in allowing us to use illustrations of their products. Similarly, a number of user companies have kindly allowed us to include photographs of their machines to illustrate effective applications of guards and safeguarding devices. We are grateful to them for their help and co-operation. Where these illustrations are used in the text, we have given suitable acknowledgement of the source.

We also gratefully acknowledge the help of the British Standards Institution in giving permission to reproduce data from European and international standards. The extracts used are reproduced under BSI's licence number 2001SK/0129. Copies of the standards referred to in the text, and those listed in Appendix 1, can be obtained from BSI Customer Services, 389, Chiswick High Road, London W4 4AL or, outside the UK, from the particular national standard-making organization.

In dealing with the subject of safety with machinery it seemed logical to go beyond the design and manufacturing aspects and include consideration of the actual use of machinery, systems of working and plant layout. In doing this there has been, inevitably, some repetition of the subject matter in different chapters but we felt it was better to do this and make the chapters 'stand alone' to ensure points were covered where they were appropriate to the text rather than rely on the reader's memory of, or need to refer to, items that were covered in other chapters.

The basic principles involved in the design of mechanical guards has changed little in recent years although its base may have been widened by the introduction of new materials and manufacturing techniques. However, the area in which the greatest changes have occurred is in the use of electronic and programmable electronic control and safety circuits. This is a rapidly developing field, advancing in parallel with developments in computer and software technologies and is spawning a growing number of standards. Some of the basic principles being developed in these emerging standards are covered in the text. However, the range and complexity of equipment currently available is so extensive and the variability of application so great that we have had to restrict ourselves to covering only the basic principles involved. In this we hope that we have given the engineer sufficient information to enable him/her to select equipment and to judge its suitability for the application. Where these complex safety systems are built into machinery designs, it is inevitable that much reliance will have to be placed on the technical knowledge of the company offering the equipment. This, in turn, places a considerable burden of responsibility on the competence and integrity of the technical representative and salesman.

Machines do not run themselves, they require operators and for the operator to work at his most efficient the machine, the type of safeguard and layout of controls and instrumentation must be such that they assist – rather than hinder – the operator. An operator who can work easily and not have to 'fight' the machine will also work more safely. The safety value of matching the operating requirements of the machine with the ability and physique of the operator is being increasingly recognized both in legislation and standards. Therefore, a chapter has been included on Ergonomics which deals with the broad issues of the operator/machine interface.

The behaviour of the operators, service engineers and others involved, also has an influence on the safe use of machinery and a chapter has been included that deals with safe working and the safety benefits that can be derived from an organization and management that has an appropriate safety culture. Many of the safe working practices discussed have become accepted as standard practices in the more technologically advanced manufacturing nations of the western world.

Similarly, the environmental conditions under which machines have to be operated is another factor in the complex formula of influences on safety in machine operations. This has been considered, not only in respect of the atmospheric aspect of the environment, but also in respect of physical layout, provision of services, access, lighting, noise, etc.

In this book we have aimed at giving the designer and engineer sufficient information to enable them to be confident that the equipment they have provided will satisfy the sometimes conflicting demands of the operator, the employer and legislation. We hope we have succeeded.

John Ridley
Dick Pearce
February 2002

Part I
Safeguarding philosophy and strategy

Ever since machinery was first developed to help man with his labours a heavy price in injuries and damage has been paid for the convenience. In the early days of the Industrial Revolution when labour was cheap, little regard was paid to the pain and suffering of injured workers. But the late 18th and the 19th centuries saw great changes in social attitudes and a growing recognition of the value of the people who worked the machines. This resulted in great strides being made in ways and means of providing protection for them. The enormous advances in technology made in more recent years have brought new hazards that have required new techniques to be developed to provide the degree of protection that society now expects employers to provide for their employees.

This part deals with some of the principles involved in providing safeguards, gives basic examples of common hazards and how they can be dealt with. It explains the process of hazard and risk reduction through the carrying out of risk assessments and how the findings can be used to determine safety integrity levels.

Chapter 1
Safeguarding of work equipment

1.1 Introduction

As machinery and plant have become more complex, so the techniques for protecting the operators have become more sophisticated. However, there are many of the older, simpler machines still in use, some of which may date from before the days when the safety of the operator was a matter for concern.

This volume is aimed at all who design, manufacture, use, maintain, modify, manage, inspect or advise on machinery, plant and component parts. It is also relevant to those who, while not directly involved with this equipment, have legal and moral responsibilites for ensuring that it is safe when put to use.

It sets out to describe the range of techniques available to the designer, works manager and engineer for guarding a whole range of machinery from the simplest to the most complex. Necessarily it covers a great many techniques and practices but to keep the text as brief as possible, the techniques and practices are descibed in outline only. Diagrams are provided where pertinent to assist in understanding the methods of operation and to enable the selection of an appropriate type of guard to be made. For further information on techniques and practices reference should be made to international and European standards, many of which are listed in Appendix 1. These standards detail the requirements that need to be met to give conformity with current health and safety legislation and hence ensure a high degree of operator safety. Conformity with a national or international standard is normally recognized as giving compliance with legislative requirements.

The text of this book applies to plant which may be manually or power operated, and extends to include equipment such as robots and pressure vessels that contain stored energy and to process plant in which the substances being processed may themselves be a hazard if they escape. The techniques described are applicable to the wide range of plant and equipment that is currently found in many workplaces and demonstrate many of the different methods by which protection of the operator can be achieved.

Over the years, individual countries have developed machinery safety standards to suit their particular methods of operating and their attitudes towards safety. However, as manufacturing has become more global so there has been, and is increasingly, a move towards international standardization. Where international standards exist, they are used as the basis for the text. Where they do not, the text reflects the best internationally accepted practices.

Ensuring that machines are provided with a suitable level of safeguarding serves two purposes. Firstly, it provides protection for the operator when using the machine. Secondly, conforming with the appropriate official published standard – whether EN, IEC or ISO – ensures that the machine complies with the conditions for importation and sale in the EU. In the latter case a technical file on the machine must be prepared of which an essential element is evidence of conformity with the appropriate standard.

Certain words that have specific health and safety meanings keep recurring throughout the book. Definitions of the meanings of those words are gathered together in a glossary in Appendix 2. Similarly, acknowledged abbreviations are used – the first time in parentheses after the full title – and these have been gathered together in Appendix 4.

Throughout the book, a variety of different methods of providing protection against machinery hazards are shown. However, these are not necessarily the only way in which the desired protection can be achieved. If a 'non-standard' method of providing protection is used the manufacturer and user may be called upon to justify their reason for using it.

It is incumbent upon the designer and engineer to select one, or a combination of more than one, method of safeguarding to suit a particular machine, method of working and safety culture. Those concerned must ensure that the method or means selected provides the level of protection that the international community is coming to expect for the people employed to operate the machines and equipment.

1.2 Design considerations

When designing machinery not only must the designer consider the efficiency of performance, achieving desired output and the economies of manufacture but he must also ensure that the finished machine will be safe in use and will not present risks of injury or damage at any stage of the machines life 'from the cradle to the grave'.

1.3 'Cradle to the grave' concept

Considerations of the safety of plant and machinery should cover every aspect of the life of the equipment including:

(a) design
(b) manufacture

(c) transport
(d) installation and erection
(e) testing
(f) commissioning and preparation for production
(g) operation from start-up to shut-down
(h) setting, adjusting and process change
(i) cleaning
(j) maintenance, repair and overhaul
(k) removing from service and dismantling
(l) disposal of parts especially if contaminated by hazardous materials.

For each of these stages consideration should be given to how the work is to be carried out and what safety features are required, remembering that it is the safety of others as well as the operator that need to be considered.

Where conflict arises between two or more safety considerations the aim must always be to reduce the overall risks as far as possible, giving greatest consideration to the aspect that poses the greatest risk.

1.4 Designer's responsibility

In the early design stages, particular aspects the designer should consider include:

1 Carrying out a preliminary design risk assessment to identify potential hazards that changes to the design could avoid. This is particularly important where new plant or equipment is to be installed in an existing work area with other plant and people movement.
2 Incorporating 'state-of-the-art' safety arrangements.
3 Utilizing feedback from other users of similar equipment to identify and eliminate hazards and improve ergonomic features.
4 Complying with legislative requirements in both the manufacturing and customer countries and ensuring that the design satisfies both these requirements.
5 Designing with the end user in mind – whether to a specific order or speculatively to offer on the open market.
6 Ensuring components, particularly those that are bought in, are compatible with the materials and other equipment with which they are likely to have contact and provide a matching degree of reliability.
7 Where control equipment is to be linked to other controls, ensuring compatibility of signals and responses.
8 Ensuring that when put into operation, the equipment does not interfere with the functioning of adjacent equipment – physically, electrically or electromagnetically.
9 The safety implications of possible changes in use and of the misuse of the equipment.

As the design progresses, other factors may arise that the designer will need to resolve.

Particular attention should be paid where access is necessary to automated or remotely controlled plant. This is especially important with robots since they retain stored energy whose release can initiate machine movement even when the control equipment is isolated.

In considering methods for providing protection, particular attention should be directed at how the machine is to be operated – what the operator should do, i.e. the safe method of operation, as opposed to what the operator wants to do (with its short cuts to increase output and boost take-home pay).

Wherever there is movement of, or in, a machine there is a potential hazard. It is incumbent on the designer to ensure that as many of the dangerous moving parts are kept within the machine frame as far as possible or contained by suitable enclosures so as not to be easily accessible. Where this is not possible, suitable safeguarding should be provided. The concept of safe by position only holds true if access to the danger zone requires the physical removal of guards or other machine parts.

In considering the safety aspects of the use of a machine the designer needs to include operations such as setting, adjustment, removal of jammed materials, preventive maintenance, lubrication, tool or component replacement, cleaning, etc.

The layout of the machine should take account of ergonomic principles matched to the required method of operation and the need for the operator to move around the machine. Lubrication, air blowing and cleaning should be either automatic or arranged to be carried out from outside the guards. The designer must be aware of and estimate possible noise levels generated by the machine and provide sound deadening insulation to guards where necessary. Any vibrations should be reduced to a minimum by ensuring rotating parts are balanced.

If a machine is later modified the designer of the modification should ensure that the modification does not reduce the level of safety of the machine. If the modification does reduce the level of safety, the user should provide additional guards or safeguards to restore to, or improve on, the original level of safety.

1.5 Safeguarding principles

The overriding safety principle to be followed in designing safeguards for machinery and plant is that it must permit safe operation without risk to the health of the operators.

The most effective and economic way of achieving this result is to incorporate the safeguards into the initial design of the machine. Thus, one of the first aims of a designer must be to design-out hazards. Factors such as speed, temperature, pressure should be reduced to the lowest level commensurate with meeting the desired performance of the machine.

However, it is recognized that much existing, and often old, machinery is still in use. When new guards are to be designed and constructed they should meet current requirements and be to modern standards.

No matter how extensive or effective a guard is, if it interferes with the operation of the machine, resulting in either a reduction of productivity and consequently the pay of the operator or frustrates the operator in his working rhythm, that guard will eventually be by-passed or removed. Consequently, it cannot be considered suitable. It is a challenge for the guard manufacturer to design and make guards that assist and improve the output from a machine while at the same time providing the required level of protection.

If an experienced operator, who generally has more knowledge of his machine than the designer, can find a faster or easier but equally safe way of working that is prevented by an existing guard, that guard needs to be modified to accommodate the improved method of working or eventually it will be by-passed or removed by the operator.

Similarly, guards must not interfere with the taught and instructed way of operating. It is not a function of the guard to protect the worker from unacceptable, unofficial or devious ways of operating the machine although the possible hazards arising from deviant operating methods should be considered in the design risk assessment. With a properly designed guard the operator should be able to work at his natural and most effective rate such that his earning potential is ensured or even enhanced.

1.6 Guarding strategy

The prime and major aim of the strategy must be the elimination of any hazards that may be faced by the operator during use of the machine. If complete elimination of hazards is not feasible, the secondary aim must be to reduce the likely ill effects of any remaining (residual) hazards to an absolute minimum by the provision of suitable safeguarding devices.

To ensure that all safety aspects are considered in the process of designing machinery or plant, a pro-forma procedure should be followed. This should comprise four stages:

1 Design hazard identification and elimination
2 Risk assessment of residual hazards
3 Risk reduction through provision of safeguards
4 Warning to users of any remaining residual operating risks

This procedure is considered in detail in Chapter 4.

Implementing this strategy requires that the designer and manufacturer have knowledge of the use to which the machine will be put so that they can recommend how it is to be operated and maintained. Possible alternative uses to which the machine may be put or for which it could be adapted should also be considered where they can be foreseen.

The safety aspects involved in each of the stages of the machine life should be considered during the design process. Design assessments of the risks involved in each of these stages should be made as the design of the machine develops.

1.7 Standard-making bodies

All countries have their own standards-making bodies for the development of national and domestic standards. On a wider front, those national bodies participate in the work of both the European and international standard-making bodies.

European and international standard-making bodies now co-operate closely in the development of standards to prevent needless duplication of effort. The relevant bodies are:

	Mechanical	Electrical
European	Comité Européen de Normalisation (CEN)	Comité Européen de Normalisation Electrotechnique (CENELEC)
International	International Standards Organisation (ISO)	International Electrotechnical Committee (IEC)

Standards promulgated by these bodies take precedence over national standards in the participating countries.

1.8 European approach to machine guarding

Within the European Union (EU), standards for the safeguarding of machinery are considered at three separate levels designated by type.

Type A standards (fundamental safety standards) giving basic concepts, principles for design and general aspects that can be applied to all machinery.

Type B standards (group safety standards) dealing with one safety aspect or one type of safety-related device that can be used across a range of machinery:
  – type B_1 standards cover particular safety aspects (e.g. safety distances, surface temperature, noise, etc.);
  – type B_2 standards cover safety-related devices (e.g. two-hand controls, interlocking devices, pressure sensitive devices, guards, etc.).

Type C standards (machine-specific standards) giving detailed safety requirements for a particular machine or group of machines.

European (EN) standards make reference to the type status of the particular document.

1.9 Unit vs holistic approach to safeguarding

Hazard identification on a machine will highlight a number of separate hazards at different places on the machine. The designer or maintenance engineer with responsibility for providing safeguarding on a machine has two basic avenues of approach. He/she can either provide separate guards for each identified hazard point – the unit approach – or enclose groups of hazard points in a simple enclosing guard – the holistic approach.

1.9.1 Unit approach

This approach involves considering each identified hazard point separately and providing each with its own unique means of protection. This method has the advantage that, should an adjustment or setting be needed at any one point, it can be dealt with without stopping the machine or affecting the other guards. If a visual check is necessary, it also allows the operator to approach the moving parts of the machine closely and safely. Perhaps its most useful application is on large machines that have discrete hazard points to which an operator may need access. However, on smaller machines this approach results in a plethora of small guards which can look unsightly and reflect just what they are likely to be – add-ons after the machine had been built. If frequent access is needed to points, the guards can be awkward with the possibility that eventually they will be left off altogether. This method of guarding can be more expensive than the holistic approach.

1.9.2 Holistic approach

With this approach, the machine is considered as a whole, taking account of the control system as well as the operating parts and considers integrated safeguarding arrangements. These allow a more comprehensive scheme of safeguarding to be developed. This approach is more cost effective, provides a higher level of protection, is cheaper and the results are more aesthetically pleasing than the unit approach. However, it may not allow the operator to approach individual parts of the machine while it is running. Integrating the mechanical and electrical safeguards gives a high degree of protection.

Chapter 2
Factors affecting the selection and effectiveness of machine safeguards

2.1 Introduction

To be effective, safeguards must provide the desired degree of protection, be acceptable to the operator, not interfere with the manner in which the operator has to operate the machine (the taught method), and not have a detrimental effect on the wage that can be earned, i.e. interfere with the operator's rate of production or working rhythm.

It is important that the user of the guard – the machine or plant operator – is kept in mind when deciding on the type and design of safeguarding device. If circumstances permit, it can be very fruitful to talk to the operators of the type, or similar types, of machines. They spend all their time on the machine and know a great deal more about operating it than a designer can possibly ever know. The contribution an operator can make in ensuring machine safety should not be underestimated. Very often, the operator will produce the answer on how the machine needs to be guarded. It is then up to the designer to put the operator's ideas and opinions into safeguarding fact.

However, it must be remembered that whether the safeguards are based on physical barriers or on trip or sensing devices, they should be designed and constructed to ensure the desired degree of protection.

In the design of guards and safeguards for machinery consideration needs to be given to some or all of the following factors.

2.2 Basic factors

In selecting the type of safeguard to suit the machinery, consideration should be given to a number of basic factors which can influence the manufacturing process and, ultimately, the level of safety performance required of the equipment. These include:

2.2.1 Safety background

The safety background is the operating environment, safety culture and corporate attitude to safeguarding machines and their operators. Employers should appreciate that their employees – the machine operators – represent a much higher investment asset than the plant or machinery itself. Within the safety background, safeguarding means should be designed to:

(i) Comply with the legislative requirements of the customer country.
(ii) Match the safety culture of the customer country or industry or, in the case of special purpose machinery, the customer itself.
(iii) Support the aims of the customer's safety policy.
(iv) Apply ergonomic principles to the finished layout.

2.2.2 Machine specification

The machine specification should describe, in outline, the expected finished product and should include any constraints on physical size, speed, temperature, etc., that have to be applied to the machine or plant. Within these constraints the design should ensure that components that are important to the safety of its operation – whether safety related or safety critical – are designed with adequate margins of safety.

The design should also take cognizance of the environment in which the machine will be operated and such factors as the juxtapositioning of other adjacent machines, potentially hostile and hazardous atmospheres and whether the machine will be exposed to the weather need to be considered.

In developing the design of the machine, the designer should give consideration to:

(i) Size, shape and layout of the machine and its juxtaposition with adjacent machines ensuring adequate access for operation, cleaning, setting, maintenance, raw material feed and finished goods take-off.
(ii) The frequency of need for access into danger zones.
(iii) Method of feeding materials, whether automatic, by machine demand, by hand or mechanically including lifting requirements for raw material feed or finished part take-off.
(iv) Setting and adjustment procedures.
(v) Access for lubrication, maintenance and cleaning.
(vi) Use of jigs and fixtures for holding workpieces.
(vii) Lighting outside and within the machine for viewing work, setting, etc.
(viii) Provision of services – electricity, air, water, etc.

Note: No dangerous part of machinery is safe by position unless it is enclosed within a machine's frame, is behind a fixed or interlocked guard or is protected by a trip device. The fact that a dangerous part cannot normally be reached is not sufficient if, when the machine is running, that part can be reached by the use of other nearby equipment such as a ladder, etc.

2.2.3 Environmental factors

Once a customer has taken delivery of a machine, it will be set to work in the environment prevailing in the customer's workplace. Many of the environmental factors can be as important to personal health and safety as to the safe working of the machine and the designer should give consideration to the following aspects:

(i) Lighting – the general standard service illuminance levels in the work area and possible stroboscopic effects.
(ii) Hygiene requirements
    – for the health protection of the operator
    – to prevent contamination of the product.
(iii) The nature of the working environment, whether it is hot or cold, kind or hostile, protected from or exposed to the weather and whether there are dusts, fumes, noise, radiations, etc.

2.2.4 Machine operation

The ultimate aim of the designer must be to ensure that when the machine is operated it is safe and does not put the operator – or anyone else – at risk of injury. The designer should be clear how the machine is to be operated and of the needs of the operator to approach various of the danger zones in the course of his/her work. The integrity of interlocking guard arrangements should be of the highest order, particularly in those areas where it is necessary for the operator to enter a hazardous zone to feed, set or adjust the machine. Operating factors that need to be considered at the design stage include:

(i) The development of safe operating procedures which the operators can be taught and be expected to follow. With old machinery these may have developed historically and become accepted but with new machinery they will be dictated by the machine maker through the operating manuals.
(ii) Design of the safeguarding method must be appropriate to the hazards and risks faced during machine operations. This should be based on the results of a risk reduction investigation (Chapter 4). Account must be taken of the need to accommodate what the operator needs to do to operate the machine.
(iii) Use of warning and delay devices at start-up where the operator at the controls cannot see all parts of the machine.
(iv) Need to operate or move the machine with the guards open. Normally this would be during setting, adjustment or clearing a jam-up (*section 6.8*).

Factors affecting the selection and effectiveness of machine safeguards

(v) Special provisions for multi-operator machines or machines with sections that can be operated independently (*section 5.4.6*).
(vi) Position and movement of parts of the machine with respect to adjacent fixed structures or fixed parts of the machine itself (*section 3.3.2 (c)* and *(d)*).
(vii) The types of materials to be worked on or handled.
(viii) Likely ejection of material being worked on – solids, liquids, fumes, dusts, etc.
(ix) Provision of services – air, lighting, water, power, etc.
(x) Use of colour and marking to identify safe and hazardous zones of the machine (*section 7.5*).

2.2.5 Administrative

Beyond the actual designing and manufacturing of the machine there are a number of factors that the supplier needs to address both to meet legislative requirements and to ensure the most effective and efficient use of the machine. While these may not be the responsibility of the designer he/she should be aware of, and contribute to, their preparation. Typically, these factors include:

(i) The preparation of maintenance and operating instruction manuals including general information about the machine.
(ii) Arrangements for the cleaning up of spillages and the removal of waste particularly if it is a contaminated or special waste.
(iii) Need to use personal protective equipment (PPE) as a back-up to the safeguards or as protection against materials in process.
(iv) The development of safe systems of work to ensure operator safety.
(v) Level of operator training needed and the competence of supervision.

2.3 Quality assurance

The continuing safe operation of a machine is an important factor in ensuring a safe place of work. But continuing safe operation requires a high degree of reliability in a machine and its safeguards. That reliability can most effectively be achieved through a conscious and ongoing effort to maintain the quality of the machine and its components during all stages of its manufacture. Establishing and maintaining a system that ensures the quality of the finished machine requires the implementation of a process that provides checks at each stage of manufacture backed by a scheme for the rapid correction of any deviations from required standard. The success of any quality assurance scheme is founded on the competence of those operating it and also on the inculcation of a suitable attitude in all involved in the manufacturing process. Any effective quality assurance scheme should begin where manufacturing begins – at the early design stages – and should be an inherent part of every subsequent manufacturing stage. Records should be kept and every stage should be documented with the results of all checks and inspections. Many customers demand manufacture to a quality assurance scheme – and this is officially recognized in the production of lifting equipment – and expect the supplier's scheme to comply with the relevant standard.

2.3.1 Quality assurance system

This must cover all aspects of the manufacturing process including:

(a) The specification of the product, which should lay down the quality standards required from both production and safety performance of the machine.
(b) The planning of the whole manufacturing process, including an organization suitable to implement the procedures necessary to achieve the required standards.
(c) The administrative arrangements to:
    (i) prepare works instruction on how the work is to be carried out;
    (ii) develop the procedures for making and keeping records of manufacture and inspections;
    (iii) develop procedures to ensure the design complies with the specification;
    (iv) train all those involved in the process in the procedures to be followed and techniques to be used;
    (v) monitor compliance with the agreed standards and manufacturing procedures.
(d) Control of design process to ensure the specification is met.
(e) The control of materiels including materials, equipment, bought in items and bought in services.
(f) The control of manufacture ensuring at each stage of the process that the specified standards are adhered to.
(g) Inspection at each stage of manufacture and of the finished product using a sampling procedure for safety-related items and individual inspection of each safety-critical item.
(h) Conditions of storage after final inspection and before delivery or use to prevent possible deterioration of performance and condition.

Details of the quality required, the standards achieved and the findings of all inspections should be recorded.

Information on the techniques and procedures to achieve these ends is contained in BS EN ISO 9000 to 9004.

2.4 Reliability

Reliability is the ability of a machine or component to perform its required function under specified conditions for a stated period of time.

It is an essential characteristic that machines and components must possess to ensure that they can continue to function properly and be capable of being operated safely and without failure for an economic period.

There are two major approaches to the ongoing reliability of a machine and its components. The first is to ensure that all parts of the machine are of a quality to ensure freedom from failure over the expected life. This assumes, for example, that bought-in items are available that will meet that criteria. Where this cannot be guaranteed, the second approach available to the designer is based on the concept that parts will wear out so the design of the machine should allow for a high degree of maintainability and, particularly, aim to allow for the easy replacement of suspect components.

Both the designer and the manufacturer have important roles to play in ensuring a high level of reliability and there are a number of techniques they can employ to this end.

The designer must anticipate, as far as is possible, those parts of the machine or its components that could be prone to failure so that in the design the designer can endeavour to:

- avoid features in the design that experiences have shown to be unreliable;
- obtain data from incident reports for similar machines and from risk assessments;
- make the design robust, keeping stresses well below the maximum levels permitted for the particular materials by using generous safety factors;
- employ duplication (or redundancy) of components where the likelihood of both components failing at the same time is greatly reduced;
- use superior quality components and bought-out items, particularly those of proven reliability, whose failure characteristics are known and whose performance is well within the operating parameters of the component or machine;
- use quality assurance schemes to ensure manufacture matches design and achieves the required quality;
- use diversity of equipment, control systems or control media to reduce the likelihood of common mode failures;
- design to a maintenance philosophy that either:
  (a) gives ease of access to components and allows a high level of maintenance and repairability or,
  (b) facilitates the replacement of safety-related and safety-critical components.

The role of the manufacturer of the machine is vital in ensuring its reliability in service. The manufacturer's attitude, which sets the scene within which manufacture takes place, must be consistent with achieving the highest level of quality of all items produced and systems must be in place to this end. Additionally the manufacturer should:

- have a manufacturing system incorporating a validated quality assurance system;
- maintain strict control on variation from design;
- inspect and test at frequent stages during the manufacturing process to ensure conformity with design and quality standards and implement a system of rapid response to correct any identified deviations;
- implement a system of strict process control and ensure it is adhered to;
- continually monitor production methods to ensure the laid down procedures and practices are followed;
- train all those concerned with the manuacturing process in the quality assurance system and the procedures to be followed;
- prepare operating and maintenance manuals that emphasize the correct operating methods and limits of use.

Reliability depends to a great extent upon effective quality control methods and the use of proven components. Maintaining contact with the machine users and obtaining feedback on their experiences can benefit the reliability of future generations of the equipment.

As soon as a machine starts operating, the components start moving, wear occurs. Provided the machine is run within the limits outlined in the specification the wear should not be significant nor adversely affect the machine's reliability. However, to ensure continuing reliable operations places obligations on the user to:

- maintain an operating environment compatible with the long life of the machine and its components;
- operate the machine within the limits of the specification;
- ensure any necessary day-to-day servicing and maintenance recommended by the maker is carried out;
- not make changes to the machine that could adversely affect its reliability or safety of operation.

2.5 Integrity

This word is often applied to safeguards and safety systems and is used somewhat indiscriminately. Dictionary definitions include *the condition of having no part or element wanting – unbroken state – material wholesomeness.* When applied to safeguards or safety systems this would appear to imply that the provision of protection is complete and without parts or functions missing and that the condition is ongoing. However, in practice, the word is used to imply a high degree of reliability as demonstrated by:

- an ability to function effectively as a whole;
- a capacity to resist interference and the adverse effects of the operating environment;
- an ability to perform the required function satisfactorily under the foreseen operating conditions for an expected period of time;
- being of a soundness of design and fitness for the purpose intended.

2.6 Validation

Where complex control systems are to be incorporated into the safeguarding arrangements for a machine as safety-related components, it is necessary that their ability to perform the required safety functions should be validated. Initially validation should be carried out on components following a procedure outlined in BS ISO 13849-2 which requires the development of a validation plan. This validation should demonstrate that the safety-related parts perform to match the characteristics outlined in the design rationale. It should also match the performance to the categories given in clause 6 of BS ISO 13849-1. Validation can be by analysis with testing to confirm the analytical findings. Further validation is necessary when individual parts of control systems are integrated into the whole control system for the machine.

2.7 Summary

By following an established strategy and employing proven design and manufacturing techniques, the machinery produced should prove to be safe to operate over its foreseen lifespan.

Chapter 3
Typical hazards of machinery

3.1 Identification

An essential element in the safe use of machinery is the identification of hazards so that action can be taken to remove them before harm or injury are caused. This applies equally to new machines where the onus is on the designer and manufacturer as well as to existing machinery where the responsibility lies with the engineer and manager of the employer.

3.2 Agents of hazards

Machinery hazards arise from a discrete number of sources – movement, energy, sharp edges, electricity, materials, physical agents and radiations. Each of these is considered below.

Condition		Hazard
3.2.1	Movement:	
	● rotation	● entanglement (rotating parts with/without projections)
		● nipping/drawing-in (gears, nips of in-running rolls, chains, belts)
		● shear (sliding parts, spoked wheels, mowing machine blades, dough mixer blades)
		● cutting (rotary knives, abrasive wheels, bacon slicers, circular saws)
	● linear sliding	● trapping, crushing (closing platens, feed tables and fixed structures)
		● shear (between adjacent machine parts, guillotines)
		● puncture (nailguns, wire stitching, stapling, sewing needles)
	● abrasives	● friction burns (rotating drums and cylinders)
		● abrasions (abrasive wheels, linishers)
	● ejection	● material (grinding debris, leaking steam, air, hydraulic oils, dusts and fumes)

Typical hazards of machinery   19

	Condition	Hazard
3.2.1	Movement (continued):	• components (process material, components in manufacture) • machine parts (overload failure, excessive speed, jam up, broken parts)
3.2.2	Energy: • stored	• air, steam or gas under pressure (pressure storage vessels and operating cylinders) • springs (actuating cylinders, robots, machining centres) • sudden release (relief valves, vessel or pipe failure) • electrical (short circuits, discharge from capacitors, static discharge) • weights and heavy parts in an elevated position (counter weights, lift cages)
3.2.3	Sharp edges	• burrs (newly cut or formed metal, swarf) • cutting blades (guillotines, loose knives, milling tools, wood and metal working tools)
3.2.4	Electricity	• shock (exposed conductors, insulation failure, no earth connection) • short circuits (fires, explosions, arc eye, burns) • overload (fires, burns)
3.2.5	Substances	• ejection from machine (leaking seals and joints) • escape (hazardous material, high pressure steam and air, flammable gases and liquids)
3.2.6	Radiations	• ionizing (nondestructive testing (NDT), X-rays, sterilizing, nuclear) • non-ionizing (ultra violet, infrared, lasers, radio frequency, induction heating)
3.2.7	Physical agents	• noise (drumming panels, metal-to-metal contacts, transformer hum) • vibration (out-of-balance shafts, percussion tools) • pressure/vacuum (tunnelling, diving, working in rarefied atmospheres) • temperature (high – drying ovens, heat treatment, and low – cold storage) • asphyxiation (confined spaces, exhaust fumes, gas leakages) • suffocation (by granular materials, powders, grain, liquids)

3.3 Hazards from parts of machinery and work equipment

Following is a list of some of the hazards associated with machinery separated into sections typified by the process functions they perform. The list is not exhaustive nor exclusive but indicates some of the typical hazards likely to occur in machinery. Included are illustrations of some of the more common hazards and of the techniques used to protect against them. For simplicity, the safeguards shown are based on the 'unit' approach of guarding each hazard separately. In the illustrations, the international convention for hazard and safety colours is followed, i.e. areas of potential hazard coloured red and safe conditions with guards in place are coloured green.

3.3.1 Rotary movement

(a) Plain shaft

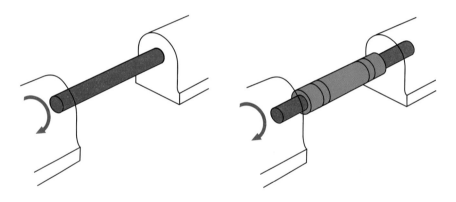

No matter how smooth the shaft, when it is rotating it will 'pick up' loose clothing and materials which will become entangled around it. An explanation of this phenomenon is given in *Appendix 4*.

Where there is no suitable place to attach a fixed guard, plain shafts can be protected by means of a loose sleeve fitted over the exposed part of the shaft with a minimum of 12mm ($\frac{1}{2}$ in) clearance. The sleeve is split along its length, placed over the shaft and the two halves bound together with circular clips or strong adhesive tapes.

(b) Shaft with projections

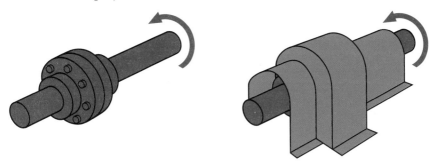

Not only will projections on rotating shafts pick up clothing and materials, they will also cause injury to any part of the body which they contact. They should be completely encased.

(c) Counter-rotating rolls

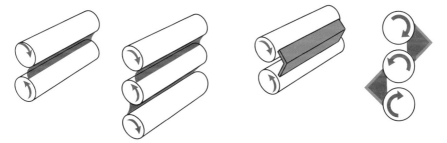

Even with considerable gaps between adjacent rolls, hands and arms can be drawn in with the risk of the whole body following. Protect with nip guards

(d) Draw rolls

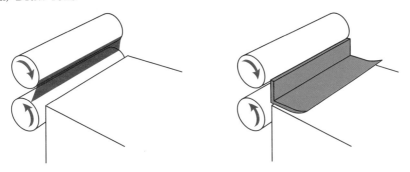

Pose a high risk since operators necessarily need to work close to them when feeding sheets and materials. Protection is provided by nip bars set to give the minimum gap for the material. Feeding is facilitated if the nip bar has a 'bird mouth' entry shape.

(e) Roller conveyor with alternate rollers driven

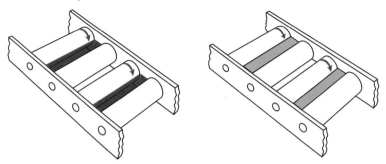

Guards should be positioned at the down stream side of the driven rollers. If all rollers are driven there is no intake hazard so guards are not needed.

(f) Axial flow fans

When included within ductwork present no hazard. The blades of free standing air movers are dangerous and should be totally enclosed by a mesh guard with the mesh large enough to allow effective air flow but small enough to prevent finger access to the moving blades.

(g) Radial flow fans

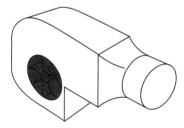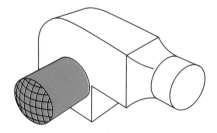

When incorporated in ductwork they present no hazard. However, the exposed inlet to the fan should be provided with a length of ducting having its inlet covered by a mesh. The length of duct and size of mesh must prevent finger or arm access to the moving impeller (*section 8.2.3*).

(h) Meshing gears

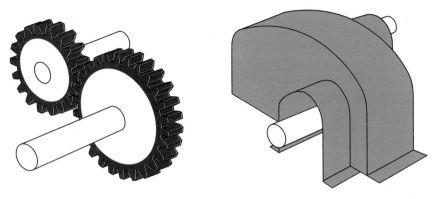

Most gears are contained within the machine frame and as such are safe by virtue of that containment. Any exposed gears must be completely encased in a fixed guard.

(i) Rotating spoked wheels

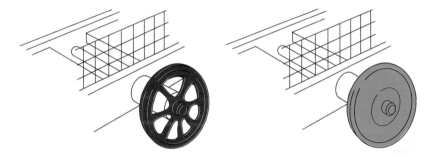

Spoked wheels for manual movement of a machine can present a hazard if attached to a rotating shaft. Protection can be provided by filling the spoked centre with a sheet metal disc. Alternately, the hand wheel can be fitted with a spring loaded dog clutch so it runs free when the shaft is rotating.

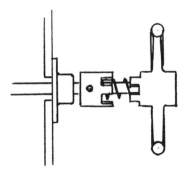

(j) Abrasive wheels

Abrasive wheels, whether fixed or hand held, should be completely shrouded except for the immediate area needed for grinding operations. The outer shroud should carry an arrow indicating direction of rotation and the maximum speed.

(k) Rotary knives

Rotary knives should be completely contained within the machine (toilet roll log cutters) or where used with manual feeding (bacon slicers) the minimum cutting edge should be exposed and be protected by a back plate. Where in situ sharpening occurs there should be suitable fire precautions especially where the process material is flammable. A special knife-holding jig should be provided for use whenever the blade is removed from the machine.

(l) Routers

On woodworking machines the part of the router blade not being used should be encased in a fixed guard. Material being worked on should be held in a jig to keep the operator's hand from the rotating blade. On hand held routers, the cutting head should be surrounded by an adjustable depth guard with provision for use with form jigs.

3.3.2 Linear movement

(a) Cutting blades

The edges of blades for cutting paper, plastics, cloth, etc., are extremely sharp and are a high risk. Wherever possible the minimum length of blade should be exposed. Paper cutting guillotines should be guarded and arranged so that the blade cannot descend until the clamp is holding the paper.

Special handling jigs should be provided for use when removing the blade for sharpening or maintenance.

Typical hazards of machinery 25

(b) Linishers

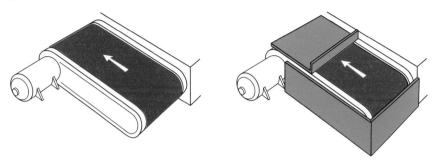

The belt should travel away from the operator and be provided with a back stop. Only the length of belt being used should be exposed. The operator end of the belt should be guarded.

(c) Trap against fixed structures

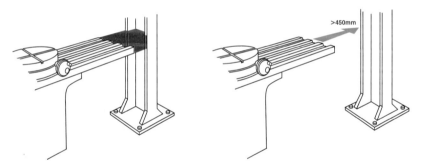

Machinery with traversing tables should be positioned so that at the extremes of movement, the ends of the table are at least 450 mm clear of any fixed structure.

(d) Trap from counter-balance weights

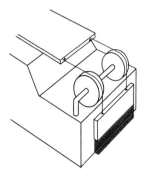 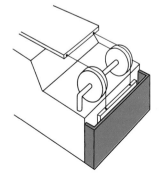

Where counter-weights are used they should be enclosed in a suitable shroud for the full length of their movement, extending to the floor or fixed parts of the machine to cover the point where a trap is created. The counter-weights of a lift should run within the lift shaft.

(e) Band saw blades

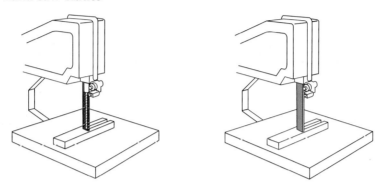

An adjustable guard should be fitted over the blade allowing exposure of only that length of blade necessary for the thickness of the material to be cut.

(f) Wire stitching and riveting machines

These machines may require the operator to hold the work close to the clinching head. Protection is provided by special trip guards that sense the presence of fingers.

(g) Scissor lifts

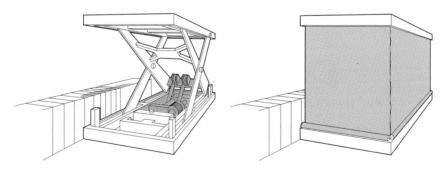

In operation, the major hazard is the shear trap formed between the edge of the table and adjoining work platforms as the table descends. Protection can be provided by roller or other blinds that prevent the operator from putting a foot under the table.

During maintenance, the major hazard is the closing of the scissor mechanism if hydraulic pressure is released while it is being worked on. Closing of the mechanism can be prevented by blocking, i.e. inserting a scotch, to prevent the lower wheels moving from the raised position. Note! under no circumstances should the hinged table be propped while working beneath it as it can tilt and allow the scissor mechanism to close.

(h) Nail guns and cartridge tools

Effectively these fire nails or studs with considerable velocity into timber, brickwork, etc. They should be fitted with a mechanism that will only allow them to be fired when the discharge port is pressed against the material into which the nail or stud is to be fixed.

(i) Fly press – closing platens and swinging counter-weight ball

Although manually operated, the platens close with considerable force. Part of that force derives from the inertia of the large counter-weight on the operating handle which can cause head injuries as the operating handle is swung.

(j) Chain saws

The main hazard is from the cutting edge of the chain and the fact that the saw can jerk when it hits a snag. Protection is provided by two-hand control, a serrated steady blade plus the wearing of special protective clothing.

(k) Lifts:
(i) in the lift shafts from movement of the lift cage and counterbalance weights particularly to those carrying out lift maintenance. Recesses should be provided in the walls of the shaft or a safe area created in the basement with stops for the lift movement
(ii) the trap with the floor sill and lintel of the openings of paternoster lifts. These should be provided with suitable hinged interlocked flap plates.

3.3.3 Combination of rotary and linear movement

(a) Rack and pinion gears

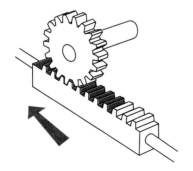
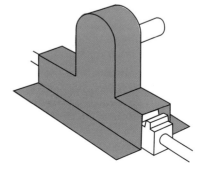

Racks and pinions gears should be enclosed in a fixed guard that completely encases both the pinion wheel and the full extent of the rack.

(b) Rolling wheels

Where there is a risk of foot injury from a wheel of a truck, traverser, etc., a 'cow-catcher' or other type of fixed guard should be fitted over the wheel. On overhead travelling cranes where access is allowed only when the power supply to the crane is isolated, guards may not be necessary. However, if work is carried out on or near the crane rails when the crane is working suitable adjustable stops should be fitted to the rail.

(c) Belt conveyor

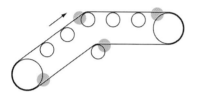 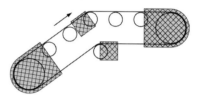

The danger points are the in-running nips where the belt changes direction and where it passes under fixed structures such as feed hoppers. These points can be guarded individually or, in the case of short conveyors, the whole side can be cased in. On long straight runs of belt conveyors trip wires, accessible to anyone trapped by a roller, provide acceptable protection (*section 6.14*).

(d) Pulley belts

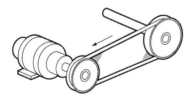 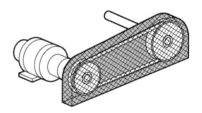

The hazards from pulley belt drives, whether flat, vee or round (rope), occur at the point where the belt runs onto the pulley. Drives of this sort also generate heat from friction slippage. Thus any guards must allow adequate ventilation otherwise the drive will overheat and fail. Weldmesh and expanded metal are suitable materials but may need a supporting frame (*section 8.2.2*). They should be positioned to ensure the belts cannot be reached by fingers. The guards of belt drives should completely surround the drive.

(e) Chains and sprockets

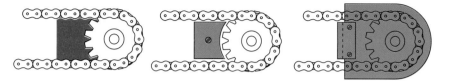

Hazards arise from both the sprocket wheel/chain intake and from the shape of the sprocket wheel itself. Guards should prevent access to both the intake and the open serrations of the sprocket wheel.

3.3.4 Contained and potential energy

(a) Robots
   Industrial robots and computer controlled machines present particular hazards through unexpected movements made in response to signals from either the controlling computer or sequence switches. They can also contain stored energy which can be released to cause unexpected movements.
   Precautions should include complete enclosure with interlocked access gates and any floor areas within the fence covered by either pressure sensitive mats or photo-electric curtains (*Figure 5.15*). Special precautions are necessary when the robot is in 'teach' mode with the teacher inside the protective fence while the robot is active. Teacher control panels should only be active through a hold-to-run switch and should include an emergency stop button.
(b) From the potential energy of counterbalance weights should the supporting chain or cable break. A shroud should be provided to cover the full length of the counter-weight's movement (*section 3.3.2.d*).
(c) Air receivers rupturing if operating pressures exceed safe working pressures.
(d) Air hoses where misdirected air jets can penetrate the skin and from the whipping movement of an unrestrained end of a flexible hose.
(e) Steam boilers and pipework where the escaping steam has a high potential heat capable of causing serious burns.
(f) The unexpected movement of pneumatic cylinders due to either:
   (i) the contained air pressure;
   (ii) the release of compressed return springs.

3.3.5 Sharp edges

(a) The cutting blades and edges of machine tools.
   Hazards arise from the sharp cutting edges of tools such as guillotine blades, bacon slicer discs, milling cutters, routers, circular and band

saw blades, band knives, doctor blades, etc., both when exposed on the machine and when being handled. On the machine the cutting edges should either be exposed only during cutting operations or the minimum length exposed for the cut to be made (*section 3.3.2.e*)

Cutting blades should be handled only when they have been secured in a special carrying jig or when the appropriate type of safety gloves are worn.

(b) Mowing machines (grass and cereal mowers, combined harvesters, etc.).

Hazards arise from the drive mechanism and shafts, especially power take-off shafts (PTOs) from the tractor, and from the scissor action of the cutters. Rotary cutters can create an additional hazard from the ejection of stones and solid materials. PTOs should be fitted with special adjustable guards. Rotary cutters should be fully shrouded (*Figure 8.4*).

(c) Hedge trimmers

These operate by a serrated cutting blade reciprocating over a series of toothed prongs. The main hazard arises from the fact that they are hand held and can be used close to the body. Protection is provided by two-hand control. Electric hedge trimmers should use either 110 V ac supply with centre tapped to earth or 220 V ac through a residual current protective device.

(d) Burrs left after machining operations.

Wherever possible machining operations should include radiusing of sharp edges. If this is not feasible, sharp edges should be removed by file or hand grinder.

3.3.6 Ejected materials or parts

(a) Abrasive wheels, in addition to the hazard from the wheels themselves, create hazards from the ejection of dusts and debris. On pedestal and bench mounted machines the debris should be ducted away and collected, care being taken to prevent ignition. With portable grinders the sparks emitted can be a fire hazard. With both, fixed and portable grinders eye protection should be used.

(b) Welding and gas burning give rise to risks of damage to eyes and burns to the operators and others nearby. Operators should wear eye protection of the appropriate obscuration. There is also a risk of igniting adjacent flammable materials and suitable fire precautions should be taken.

(c) Pipes leaking liquids can create hazards due to:
- the hazardous nature of the materials escaping requiring appropriate protective clothing to be worn when repairs are being carried out; and
- creating pools of liquids on the floor which can be a slipping hazard.

(d) Pipes leaking gases and/or fumes can create hazards through their toxic and flammable nature.

3.3.7 Leakage and escape of process material

(a) High pressure steam presents a high risk because at the point of leakage the steam may be invisible and can only be seen as a cloud of steam some way from the source of the leak. The major hazard is the high latent heat of the steam.
(b) Hazardous chemicals are a risk because of the nature of the material, which could be toxic, corrosive, carcinogenic, highly flammable or irritant.
(c) Dusts escaping from a process and not removed by the extraction equipment, can present hazards to health and the risk of fire, particularly with very fine powders, both metallic and organic. Where dust is present, the fire and explosion risk can be reduced by high standards of maintenance and housekeeping.
(d) For processes that create fumes which may escape into the atmosphere or as a direct result of the work process, i.e. colophony from soldering and NO_x from welding, suitable extraction plant should be provided.

3.3.8 Electricity

Hazards arise because it cannot be seen or heard and if felt may be too late. Precautions include adequate insulation of conductors, containment of any exposed conductors in interlocked cabinets and providing solid earthing (grounding) to all machinery and equipment. The supplies to portable equipment and tools should be either 110 V ac centre tapped to earth or, if 220 V ac, be protected by a residual current device. Provision should be made to earth or disperse static charges.

Hazards are:
- shock (ventricular fibrillation)
- fires from overloaded conductors
- explosive release of energy on short circuit
- muscular spasm from static electricity shock.

3.3.9 Temperature

Extremes of temperature can have a deleterious effect on the strength characteristics of materials and seriously impair their ability to maintain the required performance. Where machinery is liable to operate at extremes of temperature, this must be taken into account in its design. Recommended inspection frequency should be increased to suit, particularly for safety-related and safety-critical parts.

Large masses of heat such as cupolas, blast furnaces, liquid metals, etc., give effective warning of high temperatures. However, local hot spots such as arc welding, brazing, high temperature parts of a process, may have insufficient heat capacity to give obvious warning. Wherever possible, the hot components should be adequately lagged and/or suitable warning notices posted.

Contact with extemely low temperatures as well as high temperatures can cause burns so pipes and plant at these extremes of temperature should be lagged and carry warning notices.

3.3.10 Noise

Noise is recognized as a hazard that can cause deafness. Deafness prevents warning signals from being heard and can also seriously impair the quality of life. The noise generated by machinery should be assessed at the final assembly stage. Precautions, such as reducing metal to metal contact, silencing air exhausts, stiffening panels, adding sound absorbing insulation, etc., should be taken to reduce noise emissions to a minimum. The level of noise emitted by the machine should be measured and the purchaser/user informed.

Very low frequency noise can cause sympathetic vibrations in certain body organs with consequent ill health effects.

3.3.11 Vibrations

Vibrations from hand held tools such as percussion tools, vehicle steering wheels, chainsaws, fettling tools, etc., can cause a condition known as *hand-arm vibration syndrome* (HAVS) of which the most common is *vibration white finger.* The condition occurs at vibration frequencies from 2–1500 Hz and causes a narrowing of the blood vessels in the hand. Where machines or tools, that have to be hand held, generate vibrations provision should be made for special vibration resistant handles or holding devices.

3.3.12 Repetitive actions

A condition known as *work related upper limb disorder* (WRULD) can occur where the use of machinery and equipment requires the frequent forceful repetition of arm and wrist movements. It occurs in keyboarding for display screen equipment, wiring harness manufacture, chicken plucking and drawing, etc.. Equipment for use in these processes should be designed to reduce the need for repetitive actions to a minimum.

3.3.13 Radiations

Radiations, both ionizing and non-ionizing, have applications in the use of machinery and can present considerable hazards to health.

(a) Ionizing radiations can cause internal damage to body cells and interfere with the formation of new blood cells. In industrial applications, ionizing radiation sources should be contained in purpose designed capsules and the equipment in which they are installed should be arranged so that when not required for use the capsules can be withdrawn to a safe storage position (garage).

x-ray machines, in which ionizing radiations are generated electrically, should be suitably screened and arranged so they can be switched off except when irradiation is required.

For sources with a very low level of emission, such as anti-static devices and smoke alarms, the radioactive material should be fully encapsulated and may not require any further protection.

Ionizing radiations have wavelengths shorter than 10^{-8} metres with frequencies above 10^{14} kHz.

(b) Non-ionizing radiations cover a range of wavelengths from 10^{-9} up to 10^5 metres and frequencies from 1 to 10^{14} kHz. They have a range of applications:
  (i) Ultraviolet radiations, used to accelerate the drying of printing inks and in the curing of materials, can cause skin cancer and cataracts of the eye. The design of the machine should incorporate complete enclosure of the ultraviolet source to ensure there is no leakage of ultraviolet radiations.
  (ii) Visible light creates few hazards except those referred to in (c) below. Infrared, used in the control and safety circuits of machines offers few hazards except at the longer wavelength end of its spectrum.
  (iii) Lasers, used in surgery, material cutting, printing and communications, occur across the infrared, visible light and ultraviolet regions of the radiation spectrum. The high intensity beam generated can be a hazard to the eyes and in use the beam should be fully enclosed or special protective goggles worn.
  (iv) Radio frequency, used in induction heating and microwave ovens, can cause internal heating of body tissues. Machines and equipment using this technology should be provided with surrounding shielding and be securely earthed.

(c) Adequate lighting including emergency lighting, at varying levels of illuminance, is necessary for the efficient and safe use of machinery but care must be exercisd to ensure it does not create hazards. These can be caused by:
  (i) Glare from badly positioned luminaires which can interfere with the proper viewing of the work. It is a particular problem in lifting operations by crane and where inspection of the work in progress is necessary to maintain quality.
  (ii) Flicker resulting from machinery parts rotating or moving between the luminaire and the work interrupts concentration and can induce headache and migraine. The machine design should ensure there are no interposed machine parts between the lamp and the point where the light is needed.
  (iii) Shadows which prevent clear vision of work areas. The machine layout should ensure the positioning of lighting does not create shadows.
  (iv) Stroboscopic effects which occur when the gaps in rotating wheels, chucks, discs, etc., occur with the same frequency as the lighting flicker. This can create the effect of the parts being

stationary when, in fact, they are still rotating. It mainly occurs with fluorescent lamps and can be overcome either by using incandescent lamps or by arranging adjacent pairs of fluorescent lamps to be fed from power sources that are out of phase.

3.3.14 Miscellaneous

Other equipment that gives rise to hazards includes:

(a) The use of very high pressure water jetting for cutting materials and for cleaning metals, paving, etc. Hand held equipment should comprise a lance requiring two-hand control for the jet to operate. Water jet cutting fitted as part of a machine's function should be surrounded by an enclosing guard.
(b) The use of internal combustion engines in pits, tanks and other confined spaces where the engine takes the oxygen from the atmosphere and emits heavier-than-air poisonous carbon dioxide fumes. This can be fatal. Avoid if possible but if necessary copious ventilation is essential.

Chapter 4
Risk assessment, risk reduction and selection of safeguards

4.1 Introduction

One of the most effective techniques available to the designer in attempting to ensure that the machines designed are safe throughout their operational lives is the carrying out of a hazard and risk reduction exercise which includes a risk assessment. The value of this technique is officially recognized and the carrying out of a risk assessment is increasingly becoming a requirement of developing legislation. For the technique to be effective it must be carried out in a disciplined manner and follow a formal procedure.

4.2 What is a risk assessment?

A *risk assessment* is a technique for assessing the risk of harm or damage that could result from identified hazards. It is one of a progression of steps in the design process aimed at ensuring a machine is safe to use. Its findings should be used to determine the degree of protection and the type of safeguard to be provided.

The term risk assessment is sometimes used generically to refer to all the stages in the design and manufacturing process that are central to achieving its ultimate safety in use. The overall design and manufacturing process for ensuring a machine will be safe to use could more rightly be called *risk reduction*.

Two terms need to be understood:

Hazard a situation or circumstance with the potential to cause harm or damage.
Risk a compound of the probability of the hazard event occurring and the maximum extent of the likely harm.

Recognition or identification of a hazard tends to be subjective. The assessment of risk also tends to be subjective except in the few cases where statistical data are available on the reliability of equipment or components.

There are four basic stages in the design and manufacture of a machine to ensure safe use with minimum risk:

1 *Design hazard reduction* terminating with residual hazards that cannot be designed-out (ISO 12100–1; 5.2 and 5.3)
2 *Risk assessment* of the residual hazards.
3 *Risk reduction* of the residual hazards through the provision of safeguards based on the findings of the risk assessment (ISO 12100–1: 5.4).
4 Preparation of operating and maintenance manuals and information for the user (ISO 12100–1: 5.5).

At the end of the process it is unlikely that all hazards will have been removed completely, with the result that a small residual risk will remain.

When carrying out a risk assessment, a record should be made – and kept – of each hazard identified, the rationale behind the assessment of risk, the decision on the action taken and the reason for that decision. This will ensure evidence is available on the reason for a decision should any assessment be challenged at a later date when the machine is in operation.

4.3 Risk reduction strategy

When undertaking a risk reduction exercise, a laid down strategy should be followed to ensure that all the safety facets of the design process are covered. The strategy should comprise a series of discrete steps covering every stage of machine design and manufacture and consider all aspects of likely machine use. A typical strategy is shown diagrammatically in *Figure 4.1* covering the various steps from initial design to final delivery. Each step and decision taken in this procedure should be recorded, with reasons for decisions noted, against a possible later need to justify them.

The prime and major aim of the strategy must be the elimination of all hazards that may put the operator at risk. However, this may not be possible and some hazards may remain so the secondary aim must be to reduce the likely ill effects of those remaining hazards to an absolute minimum.

4.3.1 Machine life

All stages of a machine's life should be considered including hazards that could arise during:

- manufacture, whether from the size and weight of components, their physical characteristics or problems associated with securing them on machine tools;
- the size, weight and stability during transportation;

Risk assessment, risk reduction and selection of safeguards   37

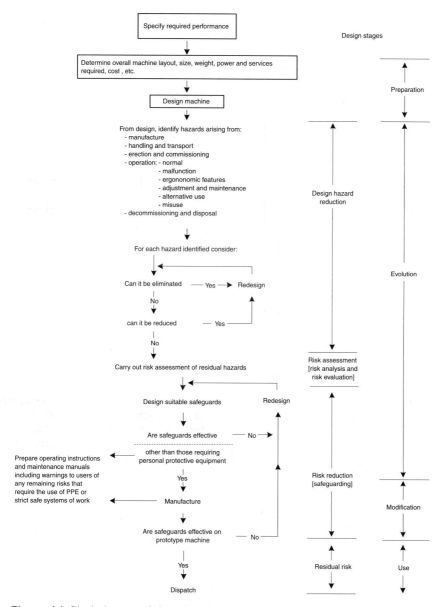

Figure 4.1 Block diagram of the risk reduction process in the design and manufacture of new and modified machines.

- the ability to erect the machine safely and to carry out any necessary commissioning tests including the possible adjustment and re-testing with guards off;
- all aspects of the operation of the machine including:
  - normal operations covered by the specification,
  - possible malfunction of components or control equipment,
  - any ergonomic hazards that may arise,
  - the ability to set, adjust and clean the machine,
  - possible alternative uses to which the machine may be put,
  - foreseeable misuse;
- the final decommissioning of the machine and its disposal taking account of hazardous materials that may have been used during its working life.

4.3.2 Hazard identification

4.3.2 (a) Hazard identification – new machinery

The safety aspects of the machine through all phases of its life must be considered. In general, the more ... complex the machinery the more sophisticated the hazard identification methods. Designer techniques for identifying hazards (EN 1050: Annex B) include:

- design risk assessment considering each function in turn and projecting possible hazards;
- analytical methods requiring detailed investigation of every function of the machine, likely failures and their effects. Specialist techniques have been developed and include:
  - 'WHAT-IF' technique for assessing the likely effects of foreseeable hazards in relatively simple machinery.
  - Preliminary Hazard Analysis (PHA) which reviews all phases of the machine life and analyses the likely hazard that can arise.
  - Fault Simulation of Control Systems (FSCS) to determine possible hazards from failures or misfunctions of the control equipment.
  - Method Organized for a Systemic Analysis of Risks (MOSAR) is a complete approach in ten steps that seeks to identify hazards and their effects.
  - Fault Tree Analysis (FTA) developing a logic diagram to trace back the cause of faults and hazards.
  - Hazard Analysis (HAZAN) analyses the possible effects of faults and hazards.
  - Failure Mode and Effect Analysis (FMEA) studies the hazardous effects of component failures.
  - Hazard and Operability Studies (HAZOP) involving multi-discipline groups who consider in detail the hazardous effects of faults.
  - The Delphi Technique is a multi-discipline approach in which the machinery is analysed in a series of progressive steps to identify hazards.

There are other equally effective custom designed systems that have the aim of identifying hazards.

Risk assessment, risk reduction and selection of safeguards 39

4.3.2 (b) Hazard identification – existing machinery

The above methods can also be employed for existing machinery but, because it is possible to carry out physical examinations, simpler methods are available that include:

- Examining accident reports. While this is a post-event procedure, action taken can prevent similar occurrences on associated machinery.
- Hazard spotting on the machinery when it is in use.
- Reports and articles in the technical press and from industry sources (federations and associations)

As the design progresses, the designer should:

- examine each of the components and parts of the machine to identify any possible hazards that they could cause. Typical of the more common hazards are considered in *Chapter 3*;
- consider hazards that might arise from specific operations such as setting, adjustment and maintenance. In this the assistance of some one versed in the use and maintenance of the type of machine in question may prove helpful;
- study records of past incidents involving similar types of machine.

This process will be assisted if the designer can draw on knowledge that extends beyond the basic operation of the machine and includes likely operating methods that may be adopted by operators and possible unconventional uses to which the machine may be put.

4.3.3 Hazard elimination and reduction

A simple procedure should:

- note each hazard identified;
- modify the design to eliminate each hazard;
- repeat these two steps until no further design improvements can be made.

4.3.4 Risk assessment

For each residual (remaining) hazard:

- analyse the cause of the hazard (hazard analysis);
- determine:
  - extent of worst likely injury;
  - the probability of an event occurring to cause that injury.

4.3.5 Risk reduction

Use the findings of the risk assessment to:

- decide on the most effective safeguard type or method;
- check the effectiveness of the safeguard against the hazard analysis;
- review and redesign the safeguarding measures until no further improvements can be made.

When satisfied that the proposed safeguarding means provides the maximum degree of protection possible, incorporate it in the design and proceed to manufacture.

Means that can be employed to reduce the risks can include:

- alternative safer materials;
- slower operating speeds;
- lower operating pressures;
- lower temperatures;
- better quality machine components of proven safety performance;
- the provision of effective safeguarding devices.

4.3.6 Monitor the effectiveness of the design

When the first, or prototype machine, has been manufactured or the modified existing equipment is put back into service, the safeguards should be checked to ensure that they protect against all identified hazards and risks. If there are any failures or questions of the adequacy of any of the measures, risk reduction procedure should be repeated and the effectiveness of the modified design rechecked.

4.3.7 Information and manuals

When the safeguarding methods have been proved adequate, operating, setting and maintenance manuals should be prepared laying down:

- the operating limits of the machine;
- the correct method for operating the machine;
- the techniques for setting and adjusting the machine;
- maintenance procedures.

4.4 Determining a Safety Integrity Level for machinery hazards

One method by which the quality of the safeguarding arrangement necessary for a machine can be determined is by reference to Safety Integrity Levels (SILs). For example, in IEC 61508 SILs are nominal numbers given to specific probability failure ranges but the concept can be extended to form the basis of a hierarchy of requirements for safety

Risk assessment, risk reduction and selection of safeguards 41

Table 4.1 Maximum permissible frequency of dangerous failure or dangerous failure rate in units of dangerous failures per hour for electrical, electronic and programmable electronic (E/E/PE) components for safeguarding circuits. (Derived from Tables 2 and 3 of IEC 61508 – 1, courtesy of British Standards Institution.)

SIL	Low demand mode of operation	High demand or continuous mode of operation
4	$\leq 10^{-5}$ to $< 10^{-4}$	$\leq 10^{-9}$ to $< 10^{-8}$
3	$\leq 10^{-4}$ to $< 10^{-3}$	$\leq 10^{-8}$ to $< 10^{-7}$
2	$\leq 10^{-3}$ to $< 10^{-2}$	$\leq 10^{-7}$ to $< 10^{-6}$
1	$\leq 10^{-2}$ to $< 10^{-1}$	$\leq 10^{-6}$ to $< 10^{-5}$

critical and safety related circuits and components. Typical values of dangerous failure rate ranges are given in *Table 4.1*.

European (EN) and international (ISO and IEC) standards refer to methods for determining the SIL but alternative pragmatic methods, which reconcile with these standards, have been developed by the authors. Essentially each of these methods of selection has, as its core, assessments of:

- consequence of the risk or severity of likely worst injury;
- the likelihood of an incident occurring.

The Pearce method introduces the concept of availability of maintenance but recognizes that there may be other human action factors which need to be considered. The Ridley method is based on quantitative values given to hazards and risks. However, the originators of particular methods may wish to introduce other factors that are seen as relevant to their particular circumstances. For example, the EN, ISO and IEC standards include consideration of the possibility of avoiding an incident.

Each of these methods arrives at a specific level of safety, in the form of an SIL, necessary to provide protection against the particular hazard or risk. A diagram of the relationship between the different methods for determining a SIL and the different safeguarding media is shown in *Figure 4.2*.

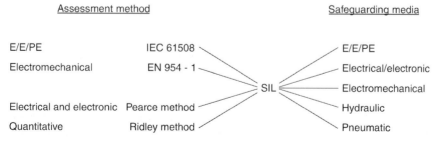

Figure 4.2 Diagram showing the relationship between assessment method and safeguarding media.

The appropriate equipment from the preferred safeguarding media should be selected to match the SIL value. Data on this are given in *Table 4.8*.

Each assessment method is considered below.

4.4.1 International standard IEC 61508

This standard is aimed at systems incorporating electric, electronic and programmable electronic (E/E/PE) equipment. Examples of assessment criteria are:

- consequence (severity of injury);
- frequency of exposure or time in hazardous zone;
- possibility of avoiding the hazardous event;
- probability of the unwanted occurrence.

In this method two of the 'consequence' categories refer to death of several people and very many people killed. This is unrealistic for manufacturing industry where failures of machinery or equipment result in, at worst, two or three fatalities. Incidents involving multiple fatalities rarely occur but when they do they are usually the result of human activities rather than the failure of a machine or one of its components.

Table 4.2 shows a method for determining the SIL using assessed criteria and *Table 4.3* gives classifications for the various criteria.

Injuries resulting from the use of machines in industry range from small cuts or abrasions requiring first aid treatment only through a range

Table 4.2 Risk graph to determine SILs for E/E/PE circuits
(Figure D2 from IEC 61508 – 5, courtesy of British Standards Institution)

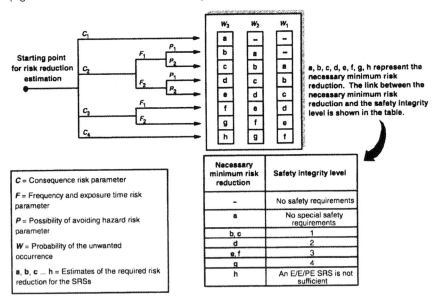

Table 4.3 Example of risk parameter classification for E/E/PE safeguarding circuits (Table D.1 from IEC 61508 – 5:1998, courtesy of British Standards Institution)

Risk parameter	Classification	Comments
Consequence (C)	C_1 Minor injury	1 The classification system has been developed to deal with injury and death of people. Other classification schemes would need to be developed for environmental or material damage
	C_2 Serious permanent injury to one or more persons; death to one person	
	C_3 Death to several people	2 For the interpretation of C_1, C_2, C_3 and C_4, the consequences of the accident and normal healing shall be taken into account
	C_4 Very many people killed	
Frequency of, and exposure time in, the hazardous zone (F)	F_1 Rare to more often exposure in the hazardous zone	3 See comment 3 above
	F_2 Frequent to permanent exposure in the hazardous zone	
Possibility of avoiding The hazardous event (P)	P_1 Possible under certain conditions	4 This parameter takes into account • operation of a process (supervised (i.e. operated by skilled or unskilled persons) or unsupervised) • rate of development of the hazardous event (for example suddenly, quickly or slowly) • ease of recognition of danger (for example seen immediately, detected by technical measures or detected without technical measures) • avoidance of hazardous event (for example escape routes possible, not possible or possible under certain circumstances) • actual safety experience (such experience may exist with an identical EUC or a similar EUC or may not exist)
	P_2 Almost impossible	

Table 4.3 (Continued)

Risk parameter	Classification	Comments
Probability of the unwanted occurrence (W)	W_1 A very slight probability that the unwanted occurrences will come to pass and only a few unwanted occurrences are likely	5 The purpose of the W factor is to estimate the frequency of the unwanted occurrence taking place without the addition of any safety-related systems (E/E/PE or other technology) but including any external risk reduction facilities
	W_2 A slight probability that the unwanted occurrences will come to pass and few unwanted occurrences are likely	6 If little or no experience exists of the EUC, or the EUC control system, or of a similar EUC and EUC control system, the estimation of the W factor may be made by calculation. In such an event a worst case prediction shall be made
	W_3 A relatively high probability that the unwanted occurrences will come to pass and frequent unwanted occurrences are likely	

of increasingly serious injuries to a fatality. The inclusion of the factors 'possibility of avoiding the hazardous event (P)' and 'frequency of, and exposure time in, the hazardous zone (F)' implies an acceptance of the need to work in the hazardous zone for periods of time and imputes an acceptance of an inability or economic reluctance to design equipment to prevent this.

The numerical value of an SIL indicates the quality of the equipment and control circuits needed to protect against the perceived hazards. Achieving that quality may rely on the use of equipment and components whose safety performance has been proven or for which measured safety performance or failure rate data are available.

4.4.2 European standard EN 945 – 1

Although this standard was developed under the aegis of CEN and is concerned with the provision of protection against machinery hazards

Table 4.4 Categories (SILs) for electrical, electromechanical and electronic safeguarding circuits. (From Figure B1 of EN 954 – 1, courtesy of British Standards Institution)

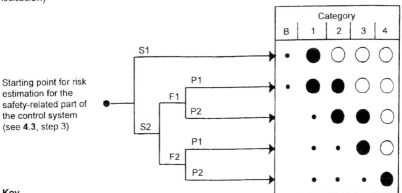

Key

S Severity of injury
S1 Slight (normally reversible) injury
S2 Serious (normally irreversible) injury including death

F Frequency and/or exposure time to the hazard
F1 Seldom to quite often and/or the exposure time is short
F2 Frequent to continuous and/or the exposure time is long

P Possibility of avoiding the hazard
P1 Possible under specific conditions
P2 Scarcely possible

Category selection
B, 1 to 4 Categories for safety-related parts of control systems
● Preferred categories for reference points (see 4.2 of the standard)
• Possible categories which can require additional measures (see **B.1** of the standard)
○ Measures which can be over dimensioned for the relevent risk

through the integrity of the safety related parts of control systems, it relies to a large extent on electro-technology and the known safety performance of the safety-related parts of electrical control systems.

The basis of the assessment of the required equipment is contained in *Table 4.4*. It again considers only two levels of injury severity, slight and serious. It also appears to condone, by inference, working in the hazardous zones.

4.4.3 The Pearce method

In this method the criterion *probability* is not considered on its own but as a combination of degree of exposure (D), operational factor (O) and maintenance factor (M) on the premise that these are the main factors considered when assessing probability. In this, the exposure factor is concerned with the extent to which an operator is exposed to danger rather than how often. It is the most significant factor in deciding the probability of an incident occurring. Operational factors are a measure of the frequency of exposure and maintenance factors influence the probability of the safeguard working when called upon to do so. The bases of the assessments to arrive at a SIL are contained in *Tables 4.5* and *4.6*. The method is directed at the operational use of the machine, not use while it is being maintained by skilled employees.

Table 4.5 Pearce risk criteria for electrical and electronic safety circuits

Risk parameter	Classification
Severity (S)	
S 1	Negligible – No disruption to normal working
S 2	Minor injury – Complete recovery within a short period
S 3	Serious injury – Complete recovery over a long period
S 4	Fatal injury or major injury – Never fully recover
Degree of exposure to hazard (D)	
D 1	Extra low – Not near machine while it is working
D 2	Low – Hand fed but not normally near dangerous parts
D 3	Medium – Hands/arms near dangerous parts when loading or removing work
D 4	High – Leaning bodily towards dangerous parts
Operational factors (O)	
O 1	Interlocking is supplementary to the use of other stopping devices
O 2	Interlocking is, or could become, the normal method of stopping the machine
Maintenance factors (M)	
M 1	Resources available for regular inspection
M 2	Resources are occasionally available for inspection
M 3	Availability of inspection resource is not reliable (breakdown maintenance is the normal practice)

Risk assessment, risk reduction and selection of safeguards 47

Table 4.6 Pearce risk graph for the allocation of SILs

Severity	Degree of exposure	Operational factors	Safety Integrity Levels (SILs) for Maintenance factors		
			M1	M2	M3
S1 — D1 — O1			1	2a	3
S1 — D1 — O2			1a	2b	3
S2 — D2 — O1			1b	3	4
S2 — D2 — O2			1c	3	4
S1 — D1 — O1					
S1 — D2 — O1			2a	4	4
S1 — D2 — O2			2b	4	4
S1 — D3 — O1			3	4	4
S1 — D1 — O2			4	4	4
S4 — D2, D3 &D4 — O1 &O2					

Severity is the dominant factor thus the probability factors D and O have less influence

See *Table 4.8* for explanation of each SIL

Note: All permutations are not shown. Those shown are the suggested solutions for the criteria chosen

4.4.4 The Ridley method

This method is a logical extension of a risk assessment and makes use of quantitative methods to arrive at a risk rating. The risk rating is a nominal figure used as a means of grading the level of protection required – the higher the risk rating the more stringent the safety measures. *Table 4.7* contains examples of typical quantitative values that can be applied to probability and severity of injury. The graph in *Figure 4.3* uses a

Table 4.7 Ridley method for developing risk rating values

Risk rating
A risk rating is a numerical value given to the risks associated with a hazard. It is a compound of the probability of an injury occurring times the likely most severe injury. Typical numerical values for probability and severity are:

Probability	Nominal numerical value	Severity	Nominal numerical value
Unlikely	1	Negligible	1
Possible	2	Minor injury	2
Probable	4	Serious injury	4
Highly probable	8	Major or multiple injury	8
Almost certain	16	Fatality	16

Risk rating = Probability × Severity

Example: Possible serious injury = 2 × 4 = 8
Almost certain minor injury = 16 × 2 = 32

---

logarithmic ordinate scale to emphasize the seriousness of increasing probability and severity.

To obtain a value for the SIL from a risk rating, on *Figure 4.3* project a horizontal line from the appropriate risk rating value on the ordinate, where it meets the appropriate curve drop a vertical line to the X axis to obtain the necessary SIL. The two curves allow selection of high or low frequency of exposure.

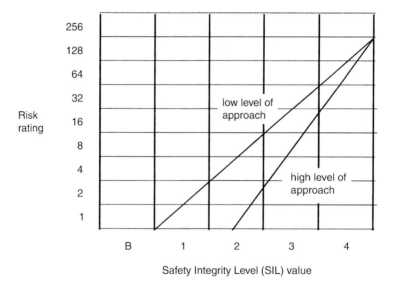

Figure 4.3 Graph for selecting the safety integrity level of protective systems using the quantitative (Ridley) method of assessment.

4.4.5 Levels of risk

Reference is often made to levels of risk as high, medium or low without relating them to particular SILs. It is difficult to specify direct relationships since there are many grey areas between the different categories of risk and between the different SILs. However, for guidance a general relationship between levels of risk and SILs is shown in *Figure 4.4*.

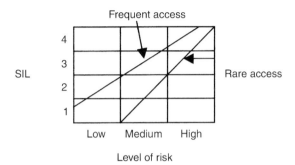

Figure 4.4 Empirical relationship between levels of risk and SILs.

When relating a level of risk to a SIL, compliance with which determines the extent of protection provided, the position selected within the band will depend on a variety of factors such as frequency of approach to danger, number of people at risk, hazardous nature of process or materials, etc. In determining the relevant SIL it is prudent to err on the side of higher risk rather than lower risk.

4.5 Selecting a safeguarding system

Selection of a safeguarding medium to protect from particular risks depends upon the type and complexity of the machine control system. Some guidance is given in IEC 61508 and EN 954 – 1 in respect of electrical and electronic safety systems, but with systems employing hydraulic or pneumatic controls it may be a question of incorporating increasing degrees of safety interlocks coupled with cross monitoring as the level of risk rises. Examples of the quality and complexity of safeguarding arrangements for different types of medium are given in the following sections.

4.5.1 Electrical, electronic and programmable electronic (E/E/PE) systems

The main requirements for these systems are contained in IEC 61508 and are summarized in *Table 4.1*. They rely on known reliability performance of the components that make up the safeguarding system.

Table 4.8 Typical electrical, electronic, hydraulic and pneumatic interlocking circuits to match SILs.

SIL	Type of interlocking	Electrical and electronic circuits	Hydraulic circuits	Pneumatic circuits
1	Single channel control interlocking	All electronic with or without software elements. (a) Hardwired with interposing devices between the interlocking device and the power switching element (Figure 9.8) (b) Hardwired with no interposing devices between the interlocking device and the power switching element (Figure 9.9)	Single channel single pneumatic interlocking valve actuating control valve (Figure 10.1)	Single channel pneumatic interlocking through interposed valves (Figure 11.2). Single channel opposed mode interlocking through interposed valves or equalizing valve (Figures 11.3 and 11.4)
	Single channel power interlocking	Including key exchange systems (Figure 9.10)	Single channel power interlocking (Figure 10.2)	
2	Single channel monitored interlocking	(a) Monitor is an audio/visual indication of failure (Figure 9.11(b)) (b) Automatic monitoring at predetermined intervals (Figure 9.11(c))	Dual channel mixed media opposed mode interlocking (Figure 10.3)	Dual channel single medium opposed mode interlocking through interposed valves (Figure 11.5)
3	Dual channel unmonitored interlocking	Includes mixed media, solenoid operated locking and dissipation of energy to enhance a single interlocking system (Figures 9.12, 9.13 and 9.14).	Dual channel mixed media opposed mode interlocking with control through an electrical safety-related control circuit (Figure 10.4)	Dual channel mixed media opposed mode interlocking through interposed valves (Figure 11.6)
4	Multi-channel monitored interlocking	Includes cross monitoring or other reliable means for detecting safety related failures and includes electronic circuits (Figure 9.15).	Dual channel mixed media opposed mode interlocking with cross-monitoring (Figure 10.5)	Dual channel mixed media opposed mode interlocking with cross-monitoring (Figure 11.7)

Table 4.9 Summary of requirements for electromechanical safety systems to match SILs. (Table 2 from EN 954 – 1, courtesy British Standards Institution.) [Note: for full requirements see the text of the standard]

Categories[1]	Summary of requirements	System behaviour[2]	Principles to achieve safety
B	Safety-related parts of control systems and/or their protective equipment, as well as their components, shall be designed, constructed, selected, assembled and combined in accordance with relevant standards so that they can withstand the expected influence.	The occurrence of a fault can lead to the loss of the safety function	Mainly characterized by selection of components.
1	Requirements of B shall apply. Well tried safety components and well tried safety principles shall be used.	The occurrence of a fault can lead to the loss of the safety function but the probability of the occurrence is lower than for category B.	"
2	Requirements of B and the use of well tried safety principles shall apply. Safety function shall be checked at suitable intervals by the machine control system.	• The occurrence of a fault can lead to the loss of the safety function between the checks. • The loss of the safety function is detected by the check.	Mainly characterized by structure.
3	Requirements of B and the use of well tried safety principles shall apply. Safety-related parts shall be designed so that: • a single fault in any of these parts does not lead to the loss of the safety function, and • whenever reasonably practicable the single fault is detected.	• When the single fault occurs the safety function is always performed. • Some but not all the faults will be detected. • Accumulation of undetected faults can lead to the loss of the safety function.	"

Table 4.9 (Continued)

Categories[1]	Summary of requirements	System behaviour[2]	Principles to achieve safety
4	Requirements of B and the use of well tried safety principles shall apply. Safety related parts shall be designed so that: • a single fault in any of these parts does not lead to the loss of the safety function, and • the single fault is detected at or before the next demand upon the safety function. If this is not possible, then an accumulation of faults shall not lead to a loss of the safety function.	• When faults occur the safety function is always performed. • The faults will be detected in time to prevent the loss of the safety function.	Mainly characterized by structure.

[1] The categories are not intended to be used in any given order or in any given hierarchy in respect of safety requirements.
[2] The risk assessment will indicate whether the total partial loss of the safety function(s) arising from faults is acceptable.

The safety circuits, which can be independent of the machine controls, generate electrical signals to the machine controls and incorporate facilities for monitoring the state of the electrical components to ensure they achieve the required condition. Where variations of state from norm result in danger, the machine stop controls are activated.

Electrical and electronic systems mainly utilize hard wired electrical circuits employing conventional electrical solenoids and contact switching in relays and interlocking switches with some electronic components in the control gear. The electrical equipment used should be a good quality, construction and of known or proven reliability.

Examples of the types of circuit – electrical, electronic, hydraulic and pneumatic – to match SILs are referred to in *Table 4.8*.

4.5.2 Electro-mechanical systems

These systems, which include hydraulic and pneumatic systems, rely on mechanically actuated switches controlling signals or power supplies to the control equipment of the machine. Their effectiveness relies on good standards of design and construction following well tried safety principles. The circuits are intended to prevent or reduce to a minimum the risks from failures of the equipment. A summary of design requirements is given in *Table 4.9*.

4.5.4 Hydraulic and pneumatic systems

Essentially these circuits function by controlling the supply of pressure fluid to the machine operating devices. The control can be direct (power interlocking) or indirect (control interlocking). Actuation of the control or interposed valves is by fluid from an interlocking valve or by a solenoid energized from an electrical interlocking switch .

Increasing degrees of protection are provided by the use of dual media and opposed mode interlocking devices plus cross-monitoring of the state of the valves. Examples of typical circuit arrangements are referred to in *Table 4.8*.

4.6 Summary

Although standards are laid down for much of the equipment and many components used in safeguarding circuits, the final selection of which components to use is likely to be made subjectively by the engineer or designer concerned. There are no hard and fast rules to dictate the type and design of circuit to be used in particular circumstances and much still depends on the personal knowledge, experience and attitude of the person carrying out the design and manufacture and on the safety culture within the workplace.

Part II
Guarding techniques

To be effective, guards and safeguarding devices must be pertinent to the application and be applied in a manner that will ensure the desired levels of protection. This part looks at the different types of guards that can be used, the techniques to make their application effective and gives typical examples. It also considers techniques of non-physical protection through the use of trips, interlocks and layout of controls. Since most machines are now powered by electricity, electrical controls and control systems of increasing complexity are being developed to match the complexity of modern machinery. Perhaps one of the most unpredictable aspects of safety in the operation of machinery is the operator. By applying ergonomic techniques to the design of safeguards the temptation to by-pass or defeat them is greatly reduced. All these facets of safeguarding are considered in this part.

Chapter 5
Mechanical guarding

5.1 Introduction

The object of a guard is to prevent access into the danger zone or to parts of a machine that could cause injury. The type of guard used will depend on the type of machine, the process and operating needs. Generally the term 'guard' refers to a fixed barrier between the operator and the dangerous part of the machine.

The particular hazards to be guarded against will be identified from the findings of the risk assessment and risk reduction investigation. A number of safeguarding techniques are available and it is incumbent upon the designer to select the one that will give the greatest protection within the parameters of the agreed operating procedures. When considering the most appropriate method of safeguarding to provide the required degree of protection, there are two main avenues of approach:

1 the provision of a guard or physical barrier, and
2 the provision of safeguarding devices that allow access into the machine but ensure that the dangerous parts are no longer dangerous when they can be reached.

This chapter reviews the various types of guard in common use. Safeguarding devices are considered in *Chapter 6*.

5.2 Guard selection

The type of guard used should be considered in the following descending order of priority:

1 a fixed guard that is securely attached to the machine;
2 an interlocked guard that, while being attached to the machine, can be moved from its safe position but is linked to the machine's power supply;
3 a movable or removable guard that can be moved from its safe position but is not linked to the machine's power supply;
4 where none of these is feasible, the use of work-holding jigs.

Safe systems of work and the provision of personal protective equipment should not be considered as primary protective means but only as back-up to guards and other safeguarding devices. The provision of guards and safeguarding devices should be backed up by training in their use. In addition, the safe methods of working that have been developed to reduce the risk of injury or damage from the machine will also maximize the effectiveness of the safeguarding provided.

5.3 Guard types

When considering the guard that will provide the required degree of protection within the operational requirements of a machine there is a range of guard types from which to make a selection (ISO 12100 – Parts 1 & 2), the most common of which are described below.

5.3.1 Fixed guard

A fixed guard is one that is held in place by devices that require a tool to release them. The securing devices can be bolts, screws, nuts, key operated deadlocks or other device acceptable to the enforcing authority. Whichever device is used it must hold the guard securely in place and not be loosened by normal machine operation or vibrations. The size of the guard will determine the number of securing devices required but the minimum number of fixings consistent with ensuring the required degree of protection should be used to encourage proper replacement after the guard has been removed for maintenance or other purposes. There should be sufficient fixings to prevent distortion of the guard when it is in the operating position. The use of captive bolts will ensure that none are lost when a guard is removed and encourages proper securing when it is replaced. To ensure the guard is not misplaced when removed from the machine, it can be attached to the machine by a hinge when it becomes a movable guard.

While welding may be used to secure the cover over an aperture to a dangerous part in a machine, it does not allow easy removal for maintenance access and, effectively, it becomes a part of the machine structure.

5.3.2 Movable guard

A movable guard is a fixed guard that can easily be moved out of position but cannot be detached from the machine. It must be designed so it can be consistently returned to its proper location. Movable guards can be hinged (*Figure 5.1*) or sliding in a supporting frame, either horizontally (*Figure 5.2*) or vertically (*Figure 5.3*). Where a movable guard needs to be secured in its operating position, the fixing should be by captive bolts or other captive securing devices. Operating cycle times can be reduced and productivity increased if movable guards are fitted with switching

58 Safety With Machinery

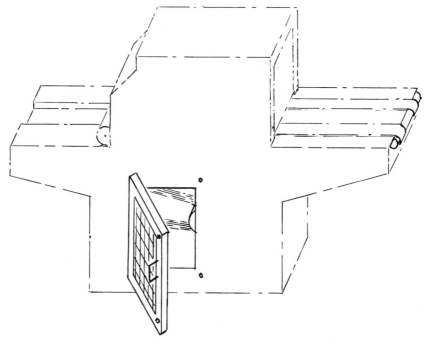

Figure 5.1 Hinged movable guard.

Figure 5.2 Horizontal sliding guard.

Mechanical guarding 59

Figure 5.3 Vertical sliding guard supported by a constant tension spring device. (Courtesy of Semi Conductor Tooling Ltd.) Note this guard is fitted to a laser welding booth and is provided with an interlocking proximity switch at its bottom right hand corner that prevents the laser firing until the guard is fully closed.

devices that are activated when the guard achieves its safe position. They then become interlocking guards.

Where the guard is moved vertically it should be provided with means to counterbalance its weight. This can be by dead weights running in containing guides or by constant tension springs as shown in *Figure 5.3*.

5.3.3 Removable guard

A removable guard is a fixed guard that can be completely removed from the machine. It must be designed so that it can, when replaced, be located properly in its correct position *(Figure 5.4)* and fixed securely to

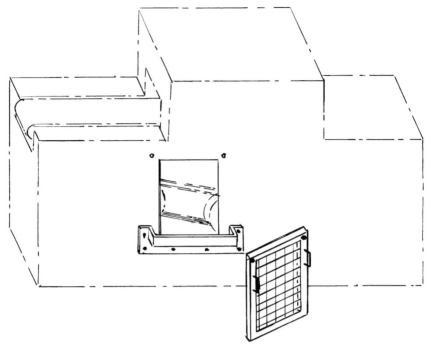

Figure 5.4 Removable guard.

the machine by captive bolts or other locking means requiring a tool to release them

As with movable guards, the number of captive bolts, or other securing devices, will depend on the size of the removable guard but should be sufficient to prevent distortion of the guard under foreseeable conditions of service. The attaching arrangement should be such that it will withstand any flexing movement or vibrations that occur in normal machine operations. Removable guards can be interlocked using two tongue operated interlocking switches (*Figure 6.5*) with one positioned at each end of the guard when the guard must be robust enough to prevent distortion or have additional securing points.

5.3.4 Adjustable guard

An adjustable guard is a fixed guard whose position as a whole can be adjusted (*Figure 5.5*) to suit the particular operation being carried out.

Once the adjustment has been made the guard elements must be firmly secured. Setting and securing these guards should be by a trained machine setter. It is essential that the movable elements are correctly positioned to prevent access to the dangerous part and that they are locked in position.

Mechanical guarding 61

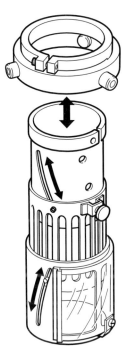

Figure 5.5 Adjustable guard where the position of the whole guard can be adjusted. (Courtesy EJA Ltd, Nelsa Machine Guarding Systems.)

5.3.5 Interlocking guard

An interlocking guard is a movable or removable guard which when the guard is moved from its safe position actuates a device that causes the machine to go to a safe state. It should be positioned to cover those parts of the machine to which immediate access is necessary as part of the normal operating process (Figures 5.6 and 5.7). With single operator

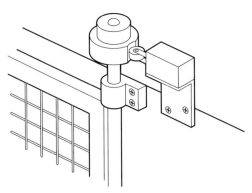

Figure 5.6 Hinged interlocking guard showing interlocking switch actuated by a cam attached to the guard hinge spindle.

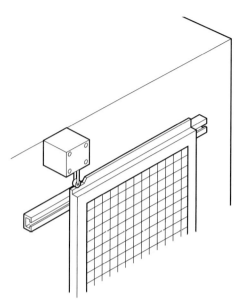

Figure 5.7 Sliding interlocking guard showing cam operated interlocking switch.

machines, where access is only necessary at the operator feed position, access to all other dangerous parts should be prevented by fixed or other suitable guards.

5.3.6 Control guard

A control guard is a guard that, when it reaches its closed or safe operating position, initiates the machine operating cycle. Its application should be restricted to cyclic processes with a short cycle time, typically single stroking press brakes, where there is a single operator only. Access to all other dangerous parts of the machine should be suitably guarded. It should not be used on continuous operation machines. The guard can be manual in operation being closed by the operator or use electrosensitive protective equipment (ESPE), such as a photo-electric curtain, when the operating cycle is initiated on clearance of either a single or double break of the curtain. In the latter, double break, case the initiation of machine movement must occur only after cessation of the second break, i.e. as the photo-electric curtain is cleared. During the subsequent operating cycle the photo-electric curtain must remain active.

5.3.7 Self adjusting guard

A self-adjusting or self-closing guard is a movable guard which covers the dangerous part of the machine but is moved by the material being worked on to allow the particular process to be carried out. It then reverts automatically to the safe position when the operation has been completed (*Figure 5.8*).

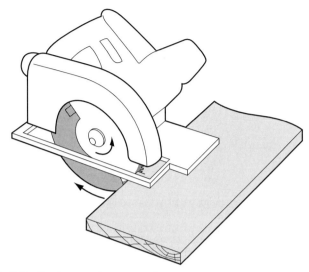

Figure 5.8, Self-adjusting guard on a wood saw.

5.3.8 Automatic guard

An automatic guard is used on single-cycle operations and is actuated by the operator striking on. This causes the guard to move to the safe position when it trips a switch initiating machine movement. The machine cannot move until the guard is in the safe position and once the machine starts moving the guard cannot be moved out of that position until the dangerous movement is completed and the machine has returned to a safe state. In its safe position the guard must prevent the operator from reaching the dangerous parts of the machine. The interlocking medium can be electrical, pneumatic or hydraulic. This type of guard is sometimes referred to as a push-away guard (*Figure 5.9*).

A similar automatic guard that is actuated by the movement of the press ram moves a barrier to a safe position preventing the operator from reaching into the danger zone. A version of the push-away guard that was used on paper cutting guillotines did not completely prevent access to the cutting blade and is no longer considered suitable. It has been superseded by photo-electric curtains (*section 6.10*).

5.3.9 Tunnel guard

A tunnel guard consists of a fixed guard in the form of a duct between the feed or take-off point and the dangerous part of a machine. It must be of a cross sectional size to accommodate the product being processed and long enough to prevent the operator reaching the dangerous part. Normally straight, in some applications that use gravity feed the tunnel can have a bend to further prevent hand access to the dangerous part (*Figure 5.10*).

64   Safety With Machinery

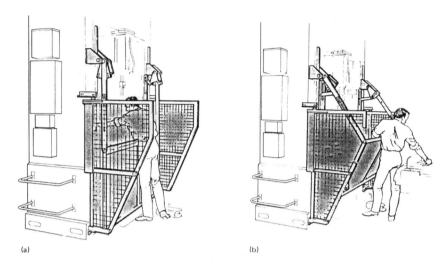

(a)   (b)

The guard, which is connected to the moving part of the press, is designed to remove the operator from the danger area before he can be trapped. (a) shows the press at top dead centre and the guard in the inner position giving free access for loading and unloading the workpiece. As the press ram descends the guard moves out to the position shown in (b) where it prevents the operator from reaching the danger area.

Figure 5.9 Automatic guard fitted to a vertical down stroking press. (Figure 51 from BS PD 5304:2000 courtesy of British Standards Institution.)

Figure 5.10 A tunnel guard at feed to a plastic granulator extended and cranked to prevent reaching to the cutters. (Courtesy of The Alan Group.)

5.3.10 Spiral guard

A spiral guard is a special purpose guard designed for application to revolving shafts to accommodate longitudinal movement along the shaft of parts of the machine. It is secured at each end to parts of the machine and can be fitted to a shaft in situ. A common application on the lead screws of lathes is shown in *Figure 5.11*.

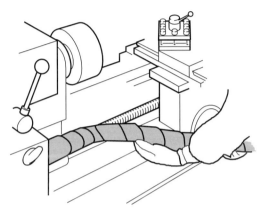

Figure 5.11 A spiral guard fitted to the lead screw of a lathe. (Courtesy EJA Ltd, Nelsa Machine Guarding Systems.)

5.3.11 Fence

The term *fence* is applied to a guard that is permanently fixed to the floor, is of a height and is set at a distance from the dangerous parts of the machine such that they cannot be reached. The fence can protect part of a machine or completely surround it. If it protects only part of the machine, other dangerous parts must be provided with suitable guards or other protective devices. Where gates are provided to give access for setting, maintenance, etc., each gate should be part of a key exchange system (*section 6.18*).

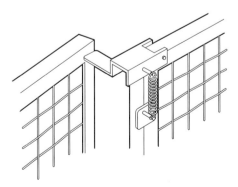

Figure 5.12 Latch to prevent interlocked gate closing inadvertently.

On lower risk machines, the gates may be interlocked by simple interlocking switches permitting quick and easy access. However, provision must be made to ensure the gates cannot inadvertently shut behind anyone. This can be achieved by the latch shown in *Figure 5.12* which will also retain the gate closed for normal operations.

5.3.12 Distance guard

Distance guard is the term used to describe a guard or fence that is set sufficiently far from the dangerous parts of a machine to preclude reaching those parts.

5.3.13 Scotch

A scotch is a mechanical restraint device that can be inserted into the machine so as to prevent physical movement of dangerous parts. It should only be possible to actuate it when the dangerous part (platen of down-stroking press or injection moulding machine) has been fully retracted. Alternatively, a progressive scotch *(Figure 5.13)* will retain the

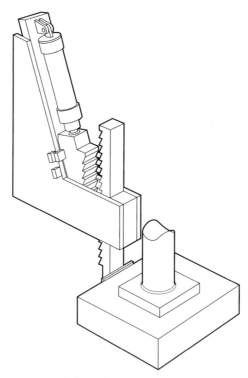

Figure 5.13 Progressive scotch for a down-stroking hydraulic press. (Courtesy Mackey Bowley International Ltd.)

Figure 5.14 Prop (scotch) under raised body of tipper lorry. (Courtesy T E Beach Ltd.)

platen at any position of its travel. Scotches can be used to retain a rise and fall guard in the raised position and should be used to prop the body of a tipper lorry in the raised position when work is being carried out on the vehicle hydraulic system or transmission (*Figure 5.14*).

5.3.14 Deterrent barrier

A deterrent barrier is, strictly, not a safety device but is intended to give warning of approach to a position or point of danger. It can consist of a handrail or fence with a gate that requires a positive action to open but is not interlocked. It should deter casual access but will not prevent determined access and as such should only be used as a back-up to other safety devices.

5.4 Other factors to consider

In the design of guards further details that need to be considered include:

1. Guards should be of robust construction and able to withstand, for the period of their working life, the conditions under which they must work.
2. The design of a guard must be acceptable to the operator, permit the full range of operating functions, not interfere with the operator's working rhythm but must accommodate the approved method of operating the machine while still providing the appropriate degree of protection.
3. A light inside the guard will facilitate viewing of the piece being worked.
4. With rotating shafts for which there is no suitable place to attach a fixed guard a loose sleeve can be used (*section 3.3.1(a)*).
5. On long machines with a series of dangerous parts along their length, such as;
   - belt conveyors, a grab wire with safety switches at each end (*section 6.14*);
   - folder-gluing and similar open processing machines, a photo-electric curtain along the length of the machine (*section 6.10*).
6. Machines with more than one operator, in addition to guards, the machine's controls should include an audible warning with built-in time delay before start up (*section 6.20*).
7. Where, for operational purposes, the machine needs to be run or moved with the guards open or the safeguarding device muted, the controls should include a hold-to-run control (*section 6.8*).
8. Machining centres on automatic sequential control should be surrounded by a fence with interlocked access gates preferably as part of a key exchange system (*sections 6.18 and 9.8.7*). Additionally a horizontal photo-electric curtain or pressure mats should cover the access floor areas inside the fence.
9. Robots should be surrounded by a fence with interlocked gates preferably as part of a key exchange system (*sections 6.18 and 9.8.7*) and horizontal photo-electric curtains or pressure mats across the accessible floor areas (*Figure 5.15*). Additional protection may be required during robot teach mode operations.
10. Equipment using power lasers must be guarded by screens that prevent leakage or beam scatter with feed opening screens being guard interlocked with the laser firing (*Figure 5.3*).
11. Total enclosure can be applied to some machines that are fully programmed and automatically fed where there is no need for the operator to approach the machine. Any openings for raw material feed and finished parts take-off should be arranged so that dangerous parts cannot be reached through them.
12. Machines with large inertia may need brakes or interlocking guards with automatic time delay guard locking (*sections 6.3 and 9.8.6*).

Mechanical guarding 69

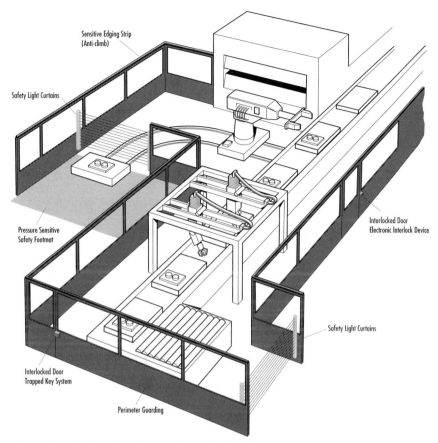

Figure 5.15 Safeguarding for an industrial robot. (Courtesy EJA Ltd. Nelsa Machine Guarding Systems.)

13 Trip devices must be positioned so that after the operator trips them he cannot reach parts while they are still dangerous (i.e. rotating or moving).
14 All machines should have an emergency stop switch (*sections 6.15* and *9.8.9*) unless it can be shown that the device will not minimize the risk.
15 Where setting or adjusting needs to be done with guards open and the machine running on power, a hold-to-run control (*section 6.8*) should be used. Only one such control should be available for use on a machine at any time.

5.5. Other techniques

In the provision of any safeguards the above types must always be considered first. However, circumstances may arise where none are technically feasible and it may be necessary to provide other methods with a lower safety integrity.

5.5.1 Jigs and fixtures

Jigs and fixtures can be used to hold the workpieces as they are being machined. This ensures that the operator's hands are kept clear of the dangerous parts of the machine during cutting operations but leaves the cutters exposed when workpieces are being re-set in the jig. With these devices, the required cutting pressure must be within the capacity of the operator to maintain. Most common applications are in the sawing and routing of timber parts.

5.5.2 Safe systems of work

In using work equipment, the machine operator should follow an agreed and trained safe system of working. To ensure the taught methods are followed requires effective operator training and extensive supervision. However, its integrity relies on human (operator) behaviour which can be unreliable and unpredictable. Safe systems of work should only be used as back-up to the use of guards and other safeguarding devices.

5.5.3 Personal protective equipment

Personal protective equipment (PPE) does not provide protection against the dangers that arise from the use of machinery but only against contamination by the material being worked on, the sharp edges of materials (burrs) and against the ill health effects of process substances in the form of dusts, fumes, liquids, etc. In respect of machinery guarding, PPE should only be used as a means to supplement other protective measures. However, PPE is essential for some processes such as welding, the cutting of metals by gas, the use of lasers or water jets, handling high or low temperature items and maintenance work on plant processing corrosives and other hazardous substances.

Chapter 6

Interlocking safeguards

6.1 Introduction

In addition to guards there are a number of techniques that can be used to provide a high degree of protection to the operator. These techniques are based on the premise that, in most cases, the hazards are related to movement of the machine and no longer exist when the machine has stopped moving.

6.2 Interlocking devices

On machinery to which movable guards have been fitted, an increased level of protection can be provided by the addition of interlocking devices that are actuated as the guard moves from the safe position. The interlocks fitted to a machine can be mechanical, electrical, electronic, hydraulic or pneumatic or a combination of two or more of these media. Duplication of devices of the same media (redundancy) improves safety integrity but is subject to the possibility of common mode failure while combinations of different media (diversity) can increase safety integrity further by reducing the possibility of common mode failures.

Where it is not feasible to fit guards, either fixed or movable, devices that detect the presence of a body or part of a body in, or approaching, the hazard zone can be used. In these cases, actuating or tripping the device causes the machine to stop or return to a safe condition.

The design of the interlocking device must be such that it cannot easily be interfered with or defeated. This can be achieved by locating it within the guard or by utilizing positive mode cam operated devices. For most operations, resetting an interlocking guard must not cause the machine to restart. Starting should only be by means of the normal start controls. However, for certain single operator single stroke short cycle operations closing an interlocking guard can be used to initiate the operating cycle (*section 5.3.6*). The operators of such processes must be suitably trained.

Interlocking devices, whether switches or valves, must have a high level of integrity and resistance to interference. Movable guards that are constrained in their movement, i.e. hinged or sliding guards, can be fitted with rotary or linear cam operated switches or valves (*section 5.3.5*).

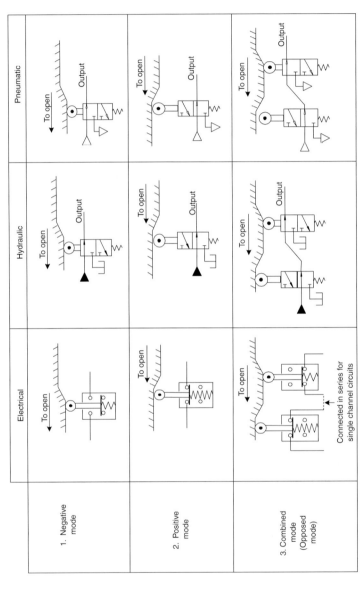

Figure 6.1 Diagram showing various modes of operation of interlocking switches or valves, all illustrations show the guard in the closed position.

Note:
1. Negative interlocking should never be used as the sole interlocking method.
2. If a single interlock only is used it should be positive mode.
3. For higher risk hydraulic and pneumatic applications, combined mode interlocking should be used. Higher risk electrical interlocking would require an additional positive mode interlock and/or monitoring.

Interlock devices can work in either *positive* or *negative* mode. These different modes of operation are shown diagrammatically in *Figure 6.1*.

In *positive* mode, the safety circuit is completed when the device is in the relaxed or non-actuated condition, i.e. for electrical switches their normally closed (nc) contacts would be made. This is achieved by providing a recess in the cam in which the cam follower can lie.

In *negative* mode, the safety circuit is completed when the cam follower is depressed, i.e. for electrical switches their normally open (no) contacts would be made. Negative mode switches or valves must not be used as the sole device in a safety-related application.

Any movement of the guard, and consequently the interlock cam, from its safe position causes the switch or valve to change state and trip the safety circuit. The effectiveness of this arrangement is dependent on the method by which the cam is secured to the guard, i.e. the cam must not be capable of adjustment or displacement.

Attachment of the actuating cams to the guard must be secure. Generally, bolts and grub screws should not be used since they can work loose with vibrations and time. On sliding guards, the linear actuating cams should be either an inherent part of the guard or welded on to the guard. For hinged guards, rotary cams can be attached to the guard by welding, or the cam plate can be attached to the hinge spindle of the guard. The attachment of the cam to the spindle should be by either:

- welding or
- spring pin, *(Figure 6.2)* or
- taper pin, *(Figure 6.3)* or
- square or spline with means to prevent linear movement such as spring pin, taper pin or circlip.

Care must be taken with the mounting of both the cam and its mating follower (part of the switch or valve) to ensure that they remain in correct alignment and in contact at all times.

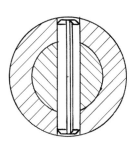

Figure 6.2 Method of securing an interlocking cam to guard hinge spindle using a spring pin.

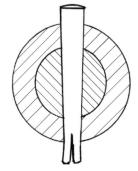

Figure 6.3 Method of securing an interlocking cam to guard hinge spindle using a taper pin. Note: the taper pin must pass right through the cam boss and have its smaller end split.

74 Safety With Machinery

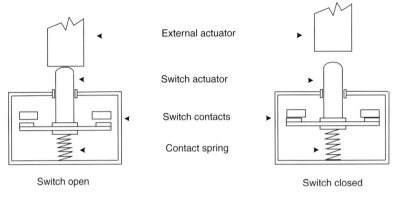

Positive switch action

Figure 6.4 Diagram of the operation of a positive break switch.

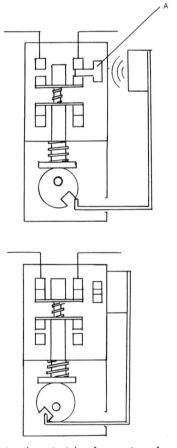

Figure 6.5 Diagram showing the principle of operation of a tongue operated interlocking switch with guard locking facility. Note: 'A' is a resonance responsive device requiring actuation before the tongue mechanism can operate.

The type of switch used in interlocking applications is important. It must be of the positive break type shown diagrammatically in *Figure 6.4* and designed so that even under fault conditions, such as spring failures or contacts welding, the contacts and hence the safety circuit can still be broken.

Tongue operated switches (*Figure 6.5*) can also be used and have the advantage of high safety integrity. The tongue should be attached to the opening or moving part of the guard with the operating body of the switch securely attached to the fixed part of the guard or machine frame. The guard must be of robust construction to ensure it does not distort and cause the tongue to lose alignment with the switch.

6.3 Guard locking

Interlocking with guard locking should be applied where it is essential that the power is isolated before the guard can be opened. With machines that have a long run-down time to stopping the guard lock should be provided with a time delay device to ensure the operator cannot reach the dangerous part before it has ceased moving. This can be a manually operated mechanical delay device (*Figure 6.6*) or a power operated bolt controlled by a timer (*Figures 6.7* and *9.17*). The medium for power locking devices can be electrical (solenoid), pneumatic or hydraulic. The guard locking arrangement must be designed so that on closing the guard, the machine cannot be started until the guard lock is fully in position. The electrical interlocking switch, timer and guard lock can be an integral unit.

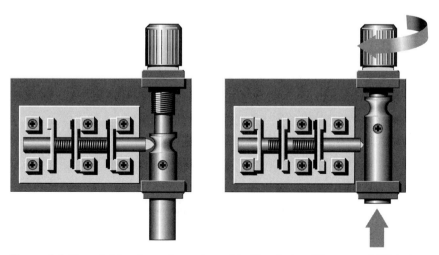

Figure 6.6 Manual delayed opening and guard locking device. (Courtesy of EJA Ltd., Guardmaster.)

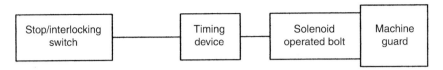

Figure 6.7 Block diagram showing the principle of operation of an electrically controlled guard locking device with timer.

6.4 Interlocking systems

The interlocking systems, whether electrical, pneumatic or hydraulic or a combination of these, can be either:

(a) *control interlocking* in which the interlocking device interacts with the machine controls causing them to bring the machine to a stop, or
(b) *power interlocking* in which the interlocking device is in the main power supply line to the machine where actuation of the interlock results in the machine being isolated from its source of power.

In both the above cases the interlocking switch or valve may be actuated by the movement of a movable guard from the safe position (EN 1088).
The interlocking device can be:

(c) a valve, either pneumatic or hydraulic;
(d) an electrical switch of the limit or positive break type (*Figure 6.4*);
(e) a magnetic switch with coded actuation and excess current protection;
(f) a diode link
(g) a high integrity electronic switch with coded signals.

6.5 Levels of risk

Typical interlocking applications for different levels of risk are:

- low risk applications
  - single channel single positive mode interlocking device *(Figures 9.8, 9.9, 10.1, 10.2 and 11.2)*
- medium risk applications
  - single channel with two opposite mode interlocking devices in series *(Figures 11.3 and 11.4)*, or
  - dual channel same medium opposed mode interlocking devices piloting separate interposed control valves or switches *(Figures 9.15a, and 11.5)*, or
  - dual channel mixed media interlocking devices, electrical in positive mode, the other in negative mode actuating separate interposed control valves or switches *(Figures 10.3 and 11.6)*

- high risk applications
  - dual channel mixed media opposite mode interlocking devices actuating separate switches or valves with cross-monitoring of the state of the interlock devices and control valves or switches (*Figures 10.5* and *11.7*) or
  - dual channel mixed media opposed mode interlocking devices with cross-monitoring connected to electronic control gear with function monitoring facilities.

Guidance on the relationship between levels of risk and SILs is given in *section 4.4.5*.

6.6 Interlocking media

Interlocking can be incorporated into mechanical, electrical, pneumatic and hydraulic control systems but the technique adopted will depend on the medium and the application. In every system care must be taken in the design to ensure that the interlocking switch or valve cannot become disconnected from its actuating cam or lever.

6.6.1 Mechanical interlocking

Mechanical interlocking by means of a direct mechanical linkage between the guard and the machine's power source must be arranged so that:

- the machine cannot start until the safety guard is closed; and
- once the machine has started its operation the guard cannot be opened until the operating cycle is complete and the machine has been brought to a halt.

Mechanical interlocking can be applied to the operating lever of an operator's manual control valve in pneumatic and hydraulic control circuits in the form of a shaped gate that prevents movement of the valve handle until the guard is closed (*Figure 6.8*).

6.6.2 Electrical interlocking

Electrical interlocking can be either:

- power interlocking (*Figures 9.10* and *9.13*); or
- control interlocking (for example *Figure 9.9*)

The integrity of electrical interlocking systems can be enhanced by the inclusion in the control circuit of monitors and electronic function checking equipment.

78 Safety With Machinery

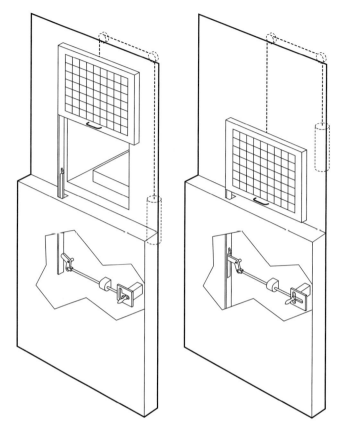

Figure 6.8 Mechanical interlocking on the control valve of a hydraulic press. (Courtesy Mackey Bowley International Ltd.)

The electrical interlocking devices can be:

(a) mechanically actuated by means of a rotary or linear cam (*Figures 5.6 and 5.7*);
(b) tongue operated switch with or without tongue locking (*Figure 6.5*);
(c) proximity switch with a matched magnetic coded transmitter and receiver;
(d) electronic switch with resonance coded transmitter and receiver.

6.6.3 Hydraulic interlocking

Hydraulic interlocking can be either:

- power interlocking in which the main hydraulic oil supply passes through a guard operated interlocking valve (*Figure 10.2 (a)* and *(b)*); or
- control interlocking in which a pilot supply from a guard actuated interlocking valve actuates a control valve or valves interposed in the main supply line to the machine cylinders (*Figures 10.1, 10.3* and *10.4*)

6.6.4 Pneumatic interlocking

Pneumatic interlocking can be either:

- power interlocking of the main air supply by a guard actuated valve (*Figure 11.4*); or
- control interlocking by pilot air from an interlocking valve actuating an interposed valve or valves in the main air supply to the machine (*Figures 11.2* and *11.3*).

6.7 Two hand controls

On single-operator machines where guarding of the feed position is not feasible, such as loose knife cutting presses, or where the consequence of an accident is potentially so serious as to warrant secondary protection, such as paper cutting guillotines, protection can be provided by a *two-hand control device* (EN 574). Essential features of these devices are:

- actuation of two separate control buttons is required to initiate a machine cycle;
- the two buttons must be actuated simultaneously;
- the two control buttons must be designed and positioned so that it is impossible to actuate both with one hand or with one hand and another part of the body;

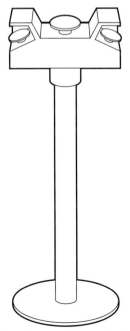

Figure 6.9 A free-standing pedestal-mounted two-hand control device incorporating an emergency stop switch.

- both buttons must be held down for the whole of the hazardous part of the operating cycle;
- release of either buttons causes the machine to revert to a safe condition;
- both buttons must be released before another stroke can be initiated.

Two-hand controls can be incorporated into the machine design or be stand-alone units (*Figure 6.9*)

6.8 Hold-to-run controls

Machines which, of necessity, need to be run with the guards open to facilitate adjustment and setting should be provided with special hold-to-run controls whose essential features are:

- only one hold-to-run control to a machine;
- when the hold-to-run control is selected all other controls are isolated;
- movement of the machine will only occur when the control button is depressed;
- on release of the control button the machine stops;
- the speed of movement of the machine is the minimum possible to carry out the setting or adjustment (about 10% or less of maximum speed);
- on large or long machines, the control can be:
  - on a single wander lead covering the length of the machine, or
  - the device can be plugged in to one of a number of local sockets when all the other local sockets must automatically become inactive, or
  - local hold-to-run control panels can be positioned around the machine where selection of any one isolates all the others;
- emergency stop switches should over-ride the hold-to-run control.

6.9 Limited movement control

Limited movement is an alternative arrangement to hold-to-run whereby a machine may be run with the guards open but moves only a limited distance for each control actuation. The limited movement should:

- require actuation of the limited movement or 'crawl' control when the work processing part of the machine should move only a predetermined minimum distance of no more than 75 mm (3 in);
- be at a minimum speed;
- have a built-in lapse time between sequential actuations of the limited movement or crawl control button.

6.10 Person sensing devices

Immaterial or intangible barriers can be applied to machines where for operational reasons the provision of physical guards is not feasible.

Known as *electro-sensitive protective equipment* (ESPE in IEC 61496 parts 1 to 4) or *personnel sensing protective equipment* (PSPE in IEC 62046). They operate by sensing the presence of a body or part of a body in the zone of detection which can cover either an area in front of the hazardous parts similar to that which would be covered by a guard or an area of floor approaching or in the hazard zone. When an object is sensed a signal is generated which interacts with the machine controls to bring the machine to a stop. These devices act as trips to stop the machine when the zone of detection is entered and as presence sensors to prevent the machine being started when the zone of detection is occupied. The zone of detection is the area over which the device is effective. Its *detection capability*, specified by the manufacturer, is the smallest size of object that will be consistently detected.

ESPE consists of three main components (*Figure 6.10*):

- a sensor that detects the presence of a body and transmits a signal to
- the control and monitor which interprets the signal, checks itself to ensure all components are functioning properly and transmits a signal to
- the main contactors in the machine supply line causing them to go to open circuit.

The ESPE can also accept and process external signals concerned with:

- *muting* (isolation) of all or some of the light beams at specific stages of a process cycle to provide access through the curtain to allow particular process functions to be performed;
- the monitoring of stopping time (*stopping performance monitoring*) where the time (measured in milli-seconds) for the machine to come to rest is a critical safety factor, i.e. within the time it would take for a person to reach the hazardous parts;

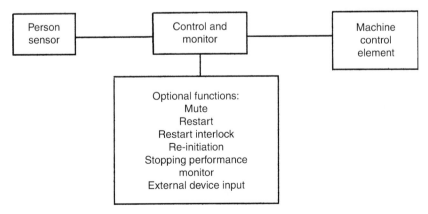

Figure 6.10 Block diagram of an electro-sensitive protective equipment (ESPE) with PSPE.

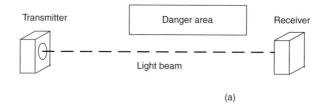

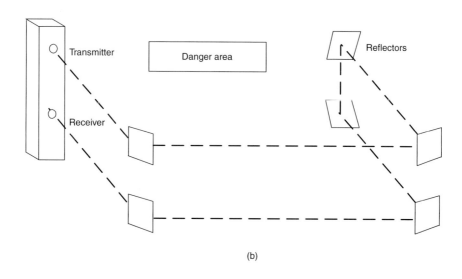

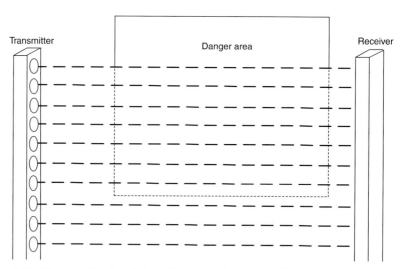

Figure 6.11 Diagram of various infrared beam configurations: (a) a single beam across an opening to dangerous parts; (b) a single beam directed by means of a series of mirrors to create a containing curtain, (c) a curtain of many beams where each beam is activated in sequence.

- tripping and other sensing devices, such as pressure sensitive mats, in applications where there is no physical barrier to prevent access to the dangerous parts.

The control equipment can contain additional internal functions concerned with restarting the machine after it has been brought to a halt by the ESPE.

The sensing medium can be photo-electric, ultrasonic, radar, laser or other electromagnetic emitter with a suitable detection capability and level of safety integrity. The most common type is the photo-electric device using infrared beams in various configurations (*Figure 6.11*)

A complete scanning cycle of the area of detection, which includes self-checking to ensure all safety functions are performing to standard, should take no longer than 12 ms. The area of detection is the zone within which a test probe (of a size specified by the manufacturer, IEC 61496 – 1 and IEC 62046) will be detected and should cover the full area of unguarded operator approach to the machine. Photo-electric devices will detect both moving and stationary objects in the zone of detection.

Mounting of the detection device can be vertical, horizontal or any angle in between but its positioning is crucial to the effectiveness of the system. The devices must be positioned sufficiently far away from the danger points to ensure that the operator cannot reach them while they are still dangerous. This must take account of the speed at which operators may approach the machine. EN 999 and IEC 62046 give advice on safety distances and orientation of detection screens for ESPEs.

There are four types of ESPE:

Type 1 specification still under consideration;
Type 2 for lesser risks where the detection of a single fault will initiate a lock-out condition;
Type 3 specification still under consideration;
Type 4 for higher risks where the detection of a single fault results in lock out.

Lock out condition occurs when various of the output signal devices of the ESPE go to open circuit, hence preventing the machine from starting until the detected faults have been cleared.

Figure 6.12 shows a schematic diagram of the interfacing of a type 4 ESPE to a machine.

Other methods of detection include the use of ultrasonic radiations and radar which detect movement but will not detect stationary or very slow moving objects. Capacitive and inductive detection methods have not yet been developed to give the safety integrity and consistency of performance necessary for use in the majority of high risk safety applications.

Close proximity sensors utilizing infrared, capacitive and optoreflective techniques have been developed to protect the fingers of operators on operations such as spot welding, stitching, sewing and riveting which require the fingers to be close to the forming tools. *Figure 6.13* shows diagrammatically the layout of these finger detecting devices.

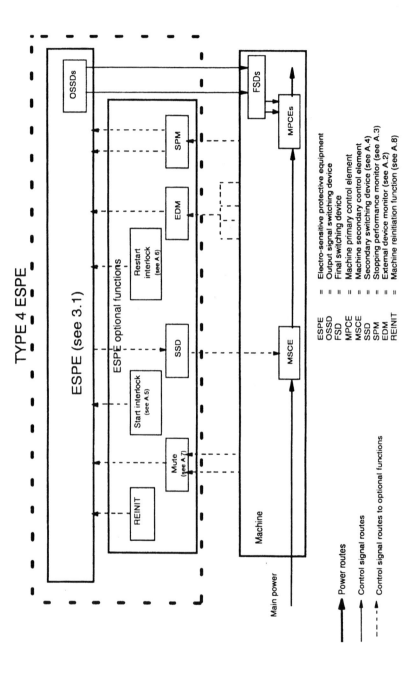

Figure 6.12 Schematic diagram of the interfacing of a Type 4 ESPE to a machine. (Figure A1 from IEC 61496 −1, courtesy of British Standards Institution.)

Interlocking safeguards 85

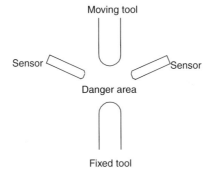

Figure 6.13 Diagram of the layout of a close proximity finger detecting device.

6.11 Lasers

Narrow beam low powered lasers can be used either as a single beam scanning an area or as a group providing protection against a discrete hazard. A single beam device can scan a work area and be programmed to recognize fixed parts of the machine while reacting to any intrusion into the area, initiating a warning signal to the control unit (*Figure 6.14*). A typical application of a scanning laser beam is the detection device fitted to the front of automatic guided vehicles (*Figure 6.15*) to sense the presence of a person or object and bring the truck to a halt before impact occurs.

Lasers can also be applied to presses used in the forming of very small components where a fixed configuration is used to detect the close approach of the operator's fingers to the forming head. In this application the laser transmitter and receiver are attached to the moving tool and provide a screen around its working edge (*Figure 6.16*). As the tool approaches the workpiece its speed is reduced. The laser protection is effective until the tool is 9 mm or less from the part to be formed when all protection is muted and the working stroke can be completed. This device is suitable only for hydraulic presses where the stopping performance is

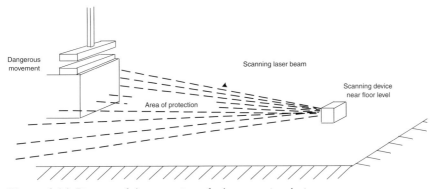

Figure 6.14 Diagram of the operation of a laser sensing device.

86 Safety With Machinery

Figure 6.15 Laser sensing device fitted to the front of an automatic guided vehicle. (Courtesy Irwin Sick Ltd.)

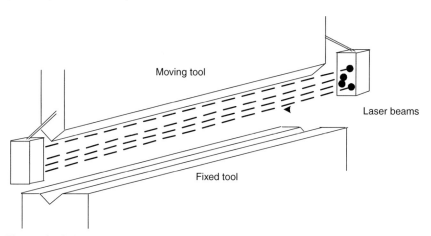

Figure 6.16 Schematic diagram of laser protection at a forming tool head. Note: the number and positioning of the beams has to be determined by calculation.

reliably predictable and unlikely to deteriorate. It is important that the power of the laser beams is below that at which damage to the eyes can occur.

6.12 Pressure sensitive mats

A pressure sensitive mat laid across the approach floor to a machine provides protection against inadvertent entry to the hazard zone (*Figure 6.17*). Stepping on that mat triggers a signal to the control equipment

Interlocking safeguards 87

Figure 6.17 Pressure sensitive mat covering the approach to dangerous parts of a machine. (Courtesy Herga Electric Ltd.)

causing the output signal switching device to go to open circuit. The mats must cover all the floor area without gaps, be arranged so they cannot be straddled and be of a material that is resistant to any contaminants present in the workplace. Pressure sensitive mats can be integrated into a safeguarding system with photo-electric equipment by feeding the signal from the mat control unit into an ESPE. The construction of the mats should conform with EN 1760 – 1 and the positioning and size of pressure sensitive mats should meet the requirements of EN 999 in respect of both approach speeds and control response time.

The operating medium of pressure sensitive mats can be either:

- electrical, consisting of two elements of conductive material held apart by compressible members. Stepping on the mat causes the conducting elements to make contact, changing the electrical characteristics of the circuit which are monitored by the control unit; or
- pneumatic, where the mat interspace has a pattern of low pressure flexible pipes which, when stepped on, cause an air pulse in the system. This pulse trips a sensitive pneumatic switch that transmits a signal to the machine controls or the ESPE; or
- fibre optics, where light is transmitted through a pattern of optical fibres laid in the mat interspace. When the mat is stepped on the optical fibres are displaced and cause a change in the light being received by a diode light sensor. This generates a signal which is transmitted to the control equipment.

In each case the integrity of the element connections must be of a high order to ensure continuity of protection.

6.13 Pressure sensitive edges and wires

A single mounting strip containing a pressure sensitive conductor can be used in applications where there is a risk of trapping between powered closing doors (*Figure 6.18*), as a trip at a feed entry for manual feeding or as bumper guards on the front of remote controlled or automatic power trucks. Deflection of the surface of the device causes changes in its electrical or optical characteristics which are monitored by the control equipment. In this they employ the same technology as pressure sensitive mats. The design of the device should ensure that it stops any movement before injury is caused. Its design should conform with EN 1760 – 2.

Roller doors

Powered doors

Figure 6.18 Diagrams of the application of pressure sensitive edge strip to prevent injury. (Courtesy EJA Ltd., Nelsa Machine Guarding Systems.)

6.14 Grab wires

On long machines, such as flat belt and troughed belt conveyors, where hazardous points occur at frequent intervals along the whole length of the machine and neither fixed guarding nor safeguarding devices such as photo-electric curtains are feasible, a degree of protection can be provided by grab wires (*Figure 6.19*). They should run along the whole length of the machine and be within reach of anyone who may be caught at any point along its length. The wire should be attached to a pull switch at each end and its length adjusted so that both switches are in their mid-position with contacts closed (*Figure 6.20*). Any movement of the wire, either pull or break, causes the switches to go to open circuit. The switches should respond to a pull on the wire of 30 N and to a movement at the centre of a wire span of not more than 150 mm. The switches should be wired in series and be part of the safety circuit. Provision should be made for supporting the wire along its length. After actuation, the machine should not be able to restart until the controls have been reset. Restarting should be by normal start controls.

An alternative to the grab wire is a pressure sensitive electric cable which, when touched, causes the machine to trip. Pressure sensitive cables have the advantage that they are static and continue to be effective in many hostile environments. They can also be carried round corners and bends without detriment to their effectiveness.

Interlocking safeguards 89

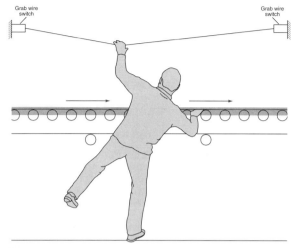

Figure 6.19 Grab wire emergency stop arrangement applied to a long conveyor.

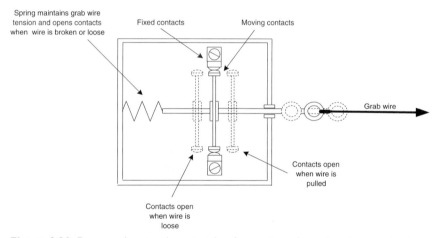

Figure 6.20 Diagram showing the principle of operation of a mid position grab wire switch

6.15 Emergency stop switches

Emergency stop switches should override all other controls to bring the machine to a stop. All machines should be provided with an emergency stop switch unless it can be shown that such a device would not contribute to minimizing the risk. Emergeny stop switches should be of the large mushroomed headed type to enable them to be actuated by various parts of the body (*Figure 6.21*). The actuating button should be coloured red. When actuated, the switch must lock in the open circuit condition and require a positive action to release it. After actuation and release of the emergency stop button the machine controls should require

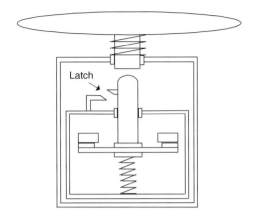

Mushroom headed latching emergency stop switch

Figure 6.21 Diagram showing the construction and principle of operation of an emergency stop switch.

resetting before the machine can be restarted using the normal start controls. The machine should not restart as the emergency stop button is released.

Emergency stop devices should be located at suitable easily accessible points around a machine and be clearly identified. They should not be used as operational stopping devices – the need to reset the controls each time an emergency switch is actuated discourages this practice. If the machine is zoned and emergency stop switches serve particular zones, they must be clearly identified with the zone they serve.

There are two categories of emergency stop switch:

Category 0
- removes source of power, or
- makes mechanical disconnection through clutch,
- if necessary includes a braking facility,

Category 1
- uses machine controls to bring the machine to a stop then reverts to category 0.

6.16 Telescopic trip switches

In the operation of certain machines, such as pillar and radial drills, where the fitting of a guard is inappropriate, protection can be provided by trip switches actuated by telescopic antennae. The switch should be mounted so that the antenna is located on the downstream side about 100–150mm from the drill and extends parallel to the drill spindle. The antenna should be extended to suit the job being carried out. Any movement of the antenna causes the switch to trip and the machine should then stop within one quarter of a revolution. The diagram in *Figure 6.22* shows a typical installation with electric braking.

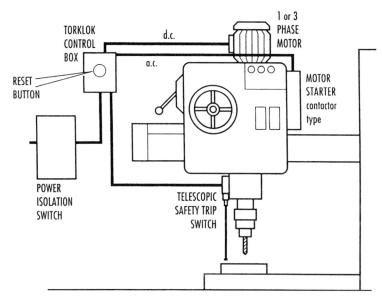

Figure 6.22 Schematic diagram of the installation of a telescopic trip switch. (Courtesy of EJA Ltd., Nelsa Machine Guarding Systems.)

6.17 Proximity switches

Proximity switches operate through the juxtapositioning of two associated parts, one containing the operating (switching) mechanism and the other the actuating means. There are no physical contacts between the parts and the devices can be tolerant of a reasonable degree of misalignment. Each part can be hermetically sealed and be resistant to a wide range of operating environments. They are particularly suitable for use where regular washing down of equipment is necessary, i.e. in the pharmaceutical and food industries.

The two main types of proximity switch are magnetic and electronic. In magnetic switches, the actuator comprises a series of magnets to form a code to which the sensor reacts. The contactors in the sensor should be generously rated with the moving contactor mounted on a leaf spring. There should be built-in over-current protection and external circuits should incorporate a fast-break fuse of a slightly lower rating.

Electronic proximity switches employ an electronically coded link between the sensor unit and the actuator. For high risk applications the sensor unit can include self-checking facilities and be incorporated into high integrity safety circuits.

6.18 Key exchange systems

On large machines having a number of guards or gates that may need to be accessed during maintenance, setting, etc., key exchange systems

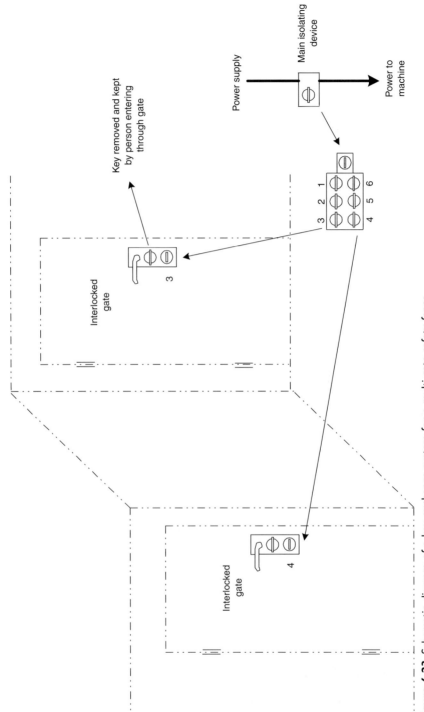

Figure 6.23 Schematic diagram of a key exchange system for a multi-gate safety fence.

provide a high level of protection to prevent inadvertent start-up of the machine while someone is still within the safety fence. The basic system operates in three stages. First, the power is turned off by the control key which is then removed from the main power switch and inserted in the secondary key board to release the secondary keys. Removal of one or more secondary keys locks the control key in the secondary board and prevents power from being switched back on. The secondary key, in turn, has to be inserted in the gate lock before the primary (gate) key can be removed and the gate opened. Removal of the primary key locks the secondary key in the gate lock and it can only be released when the primary key is replaced. The operator should keep the primary key on his person until the work has been completed when the gate can be closed and locked. The full reverse procedure has to completed before the power can be reconnected to the machine.

Each key should be coded and be unique to a lock and parking socket. The keys should only be capable of being used in the locks to which they are assigned

A schematic diagram of a key exchange system is shown in *Figure 6.23*.

6.19 Key interlock switches

Guards can be interlocked manually with a key which can be either loose or captive and which, in releasing the guard lock, also trips the machine controls. It is essential to ensure that the stopping time of the machine is less than the access time.

Variations of this type of arrangement include:

- a captive key device in which the key is attached to the movable guard and arranged so that on closing the guard it locates with the switch part. The first part of the movement of the key or handle mechanically locks the guard shut and the second part of the movement closes the contacts of the interlocking switch. The switch part of a captive key device should be located within the machine frame so it is not easily accessible. The key part of the device should be coded to match uniquely the mating switch part;
- a key switch provided with a separate lock in its actuating handle allowing actuation only when the lock is released. This should ensure that access is restricted only to those issued with a key. The issue of loose keys to employees should be strictly controlled.

For machines with a finite run-down time the interlock circuit should include a time delay guard lock.

A typical application for a captive key switch is shown in *Figure 6.24*.

6.20 Delayed start

On large machines with more than one operator where the person at the controls cannot see all the other operators, the controls should

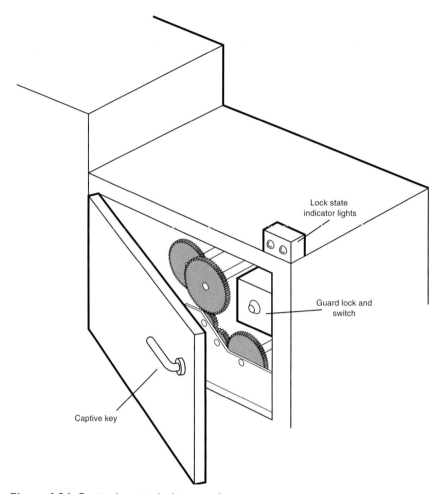

Figure 6.24 Captive key interlocking guard.

incorporate a delayed start function. When the start control is actuated an audible warning is sounded followed by a pause of at least 15 seconds before the machine starts moving. This warning and delay allows other operators to get clear of the machine or to activate an emergency stop button.

6.21 Other interlocking devices

Limiting devices should be incorporated in the machine design to ensure that the machine operates within defined constraints of movement, pressure, temperature, speed, time, etc. This reduces the likelihood of a failure due to overload, fatigue or excessive stress.

Chapter 7

Ergonomic aspects of machinery safeguarding

7.1 Introduction

Ergonomics can be defined as the study of man in relation to his working environment and work equipment. Through the application of practical techniques, ergonomics can make the machine operator's efforts more effective and his working life easier. An increasing number of European and international standards contain requirements to follow ergonomic principles and this is a factor that must be considered in the design of machines and their safeguarding arrangements.

With machinery safeguarding, as with many others matters involving a human interface, the easier it is to use the more likely it is to be used. Guards that are difficult to use or interfere with the operator's control of the machine can cause frustration and may stimulate the development of ways to by-pass the annoyance. It is incumbent on the designer and manufacturing engineer to ensure that the layout of the machine, its controls and the safeguards follow the best ergonomic practices.

7.2 Physiology

There are considerable differences in the physical build-up of different nations and ethnic groups and account should be taken of the physiological characteristics of the populace of the ultimate customer country when designing machinery and its associated safeguards. There is a great body of reference statistical data on this subject.

For muscles to work effectively they need to have a good supply of oxygen which is delivered to them by the blood circulating around the body. The quantity flowing into the arms and hands will be determined by their position relative to the heart. Below the level of the heart, the lower the position the greater the flow of blood. Above the level of the heart, the higher the limb the less the flow of blood. Ideally, the position of hands and arms when operating safeguards and controls should be between waist and shoulder levels.

Similarly, muscular movement causes blood to flow whereas static positions tend to restrict the blood flow. Wherever possible in the operation of machines the holding of controls or safeguarding devices in one position for any length of time should be avoided. Removable guards should be designed so that they can be lifted into a channel or slot – to locate and take the weight – while the securing means are being applied. Where handles are fitted to assist replacement they should be in such a position on the guard that a natural posture can be adopted for the guard's replacement, but they should be above the centre of gravity of the guard otherwise it will be unstable when lifted. The weight of a removable guard must be within the lifting capacity of the operator and be as light as possible commensurate with providing the level of safety required.

Body movements should follow natural physiological paths avoiding repetitive and body twisting movements. Where loads are required to be lifted or moved, mechanical assistance should be provided. Lifting while seated should be avoided if possible, if not the forces and movement required should be within the accepted ergonomic limits for a seated person.

If it is necessary to view work or operations through a guard in which an aperture is provided, the aperture should be at a level where it can be used naturally and does not require the operator to stoop or stretch upwards since both of these actions are potentially tiring and stressful.

By the same token, the feed tables of hand-fed machines should be about waist height. Where a number of operators on different shifts are employed at a feed station, it may be necessary to provide a low platform for the shorter operators to stand on.

Wherever possible, the operator's stance when operating a machine should be natural with awkward postures being avoided.

7.3 Controls

Controls are a major component of the interface between the operator and the machine. Their design is an important factor in the safe and effective operation of the machine. Kroemer and Grandjean[1] consider that:

> Controls constitute the 'feed forward' second 'interface' between human and machine. We may distinguish between the following:
>
> 1 Controls that require little manual effort; *push buttons, toggle switches, small hand levers, rotating and bar knobs, all of which can be operated with the fingers.*
> 2 Controls that require muscular effort; *hand wheels, cranks, heavy levers and pedals.*
>
> The right choice and arrangement of controls is essential if machines and equipment are to be operated correctly.
> The following guidelines should be followed:
>
> 1 Controls should take account of the anatomy and functioning of the limbs. *Fingers and hands should be used for quick, precise movements; arms and feet used for operations requiring force.*

2 Hand operated controls should be easily reached and grasped, be between elbow and shoulder height and be in full view.
3 Distance between controls must take account of human anatomy. *Two knobs or switches operated by the fingers should be not less than 25 mm apart; controls operated by the whole hand need to be at least 50mm apart.*
4 Push buttons, toggle switches and rotating knobs are suitable for operations needing little movement or muscular effort, small travel and high precision, and for either continuous or stepped operation (click stops).
5 Long radius levers, cranks, hand wheels and pedals are suitable for operations requiring muscular effort over a long travel with comparatively little precision.

Controls should offer some resistance to movement so they cannot be inadvertently actuated.

The direction of movement of a control should be consistent with its effect, i.e. turning clockwise to increase and anticlockwise to decrease – there are notable exceptions to this such as a valve where a clockwise movement reduces the flow, i.e. shuts the valve. With lever controls, pushing should cause a movement away from or lowering down of the part and a pulling movement of the lever control should cause a movement towards the operator or the part to rise (EN IEC 61310 -3). Controls that are likely to be operated by touch should have a distinct tactile identity (EN IEC 61310–1).

Gauges and instruments indicating the effect brought about by the movement of a control should be mounted in clear view, at eye level, and have indications that are consistent with the effect measured. They must not interfere with the operator's view of the machine.

Pedal controls should be arranged so that the operator does not have to stand on one leg to actuate them. If the pedal is above floor level, a small platform should be provided to bring the operator's feet up to the level of the pedal and arranged so that the pedal can be operated by either foot. Small movements of pedals can be through pivotal movement of the ankle with the outer edge of the foot resting on a support. Larger pedal movements require movement of the whole leg from the hip.

Operational controls should be positioned to be conveniently available to the operator at his normal machine operating position and not require difficult or awkward movements to reach them. The most frequently used switches or control levers should be located in the most accessible positions.

There should be ample space between the individual switches to prevent inadvertent actuation of an adjacent switch. Start switches should be shrouded, recessed, gated or otherwise arranged so they cannot be actuated inadvertently. Stop controls should be positioned adjacent to start controls and project or protrude above the control panel surface, be red in colour and the control circuit arranged so that stop controls override start and other controls. All switches and devices used for controlling the machine should be clearly identified, either by words or symbols, such that the operator can have no doubt of the effect of the control being operated (IEC 60204 Part 1).

Emergency stopping devices should be located at each control position and at easily accessible strategic points around the machine. Their location should be clearly identified. Emergency stop switches should have large mushroom or palm headed buttons, be coloured red and be such that they lock in when actuated (*Figure 6.21*). They should be capable of being actuated by hand, foot or other convenient part of the body. Release or reset of an emergency stop button should require a positive action but must not cause the machine to start.

The movements of control actuators should be compatible with natural human movements and their effect should be similar on similar machines.

7.4 Machine layout

In the layout of the machine, movable and removable guards should be located at the positions at which the operator requires access. If the operating procedure requires reaching through the guard with the machine running provision must be made to ensure dangerous parts cannot be reached (*Table 8.3*). When, as part of feeding, adjusting or servicing a machine, it is necessary to place parts of the body between two moving components, arrangements must ensure that the components cannot close sufficiently to cause harm (*Table 8.5*).

7.5 Colours

Colours on machines and in the workplace should follow the current international protocol:

- red — danger, emergency, stop
- green — safety, normal, go
- yellow — warning, abnormal
- blue — mandatory, obey

The use of colours on controls should follow EN IEC 60204–1, cl 10.2.

Colours can be used to attract attention, i.e. red emergency stop buttons on a yellow background; a spiral line on a rotating disc or shaft to warn of movement, and similar colours to identify related components, controls or work areas.

Certain colours also have illusory effects:

Colour		Illusory effect	
Blue and green	Cold	Makes objects look more distant	Restful
Orange and yellow	Warm	Makes items look nearer	Stimulating
Brown and buff	Neutral	Makes items appear much closer	Restful

For aesthetic reasons, the exterior of the machine, including the guards, may be in a colour consistent with the decoration of the rest of the plant so that in the safe working condition the machine presents a unified appearance. The area behind the guards where the hazard exists should be painted in a warning colour (black and yellow diagonal stripes) or a danger colour (red) so that if a guard is removed, its absence can be seen immediately.

Care must be taken in the use of colours so that they do not confuse. While contrasting colours can highlight items, too many contrasts can confuse and be counter-productive. Fluorescent colours should be avoided. Pipe lines and their contents can be identified by colour coded markings (Appendix 5).

7.6 Lighting

Levels of lighting round a machine should enable the operator to carry out his duties without suffering eye strain. The more detailed the work the higher the lighting intensity. Typical intensity values to provide adequate levels of illumination are:

Location	Standard service illuminance (lux)
Plant rooms, storage areas	150–200
Rough machining and assembly, control rooms, weaving sheds, cold strip mills	300–400
Medium machinery and assembly	500
Machines with display screen equipment	300–500
Inspection areas	750
Fine machining and assembly	1000
Fine hand work and machinery inspection	1500

Where the power supply is ac, adjacent lamps should be connected to different power supply phases to prevent a stroboscopic effect with rotating parts.

For general lighting, single point luminaires should be avoided since they create glare and shadows, fluorescent luminaires should have adjacent tubes wired out of phase.

Luminaires should be positioned where they do not create glare or deep shadow nor strong contrast in viewing areas. Viewing areas, such as display screens monitors with dark surfaces, should not be located in front of sources of bright light such as windows. For very detailed work, local lighting may be needed.

Flickering fluorescent tubes should be replaced as soon as possible since they can cause headaches, and can bring on migraines and epilectic fits.

7.7 Noise

Sound is necessary for communication, warning, recognition, etc. Noise is unwanted sound. Noise can cause a number of reactions in individuals ranging from fatigue through stress to complete disorientation. Levels above 85 dBA are potential harmful and above 90 dBA can result in noise induced hearing loss. If noise from a machine interferes with normal conversation it should be measured and action taken to contain or reduce it.

Noise reduction measures can include:

- reducing the number of metal to metal contacts by the increased use of plastics in gears, bushes, etc.;
- replacing worn bearings, gears, linkages, etc.;
- brace sheet metal parts to prevent drumming;
- provide sound absorbing linings;
- silence air exhausts;
- contain noisy equipment, such as compressors, vacuum pumps, etc., in soundproof enclosures or remove to a separate soundproof room;
- provide soundproof control rooms for operators;
- provide silencers in ducting.

Good maintenance is important in controlling noise levels.

7.8 Vibrations

Vibrations emitted by a machine waste energy and can have serious effects on the operators. Low frequencies up to 100 Hz can cause resonance in certain organs, particularly the eyes interfering with vision. At higher frequencies the effect tends to be more local to points of contact, mostly the hands, such as in the holding of angle grinders, lorry steering wheels, pneumatic drills, chain saws, etc., resulting in a condition known as vibration white finger. Vibrations can be transmitted through the foundations and structures of buildings as well as by air pulses.

Vibrations should be reduced to a minimum by balancing rotating parts and reducing the mass of reciprocating parts. Machines that generate vibrations such as reciprocating engines, compressors, etc., should be mounted on special vibration proof foundations or pads. Hand-held tools should be provided with vibration proof handles.

7.9 Rate of working

Considerable stress and fatigue can occur in workers when they are obliged to work at rates outside their natural aptitude range. Machines that dictate the rate of work should include arrangements to enable the speed to be adjusted to suit the operator's natural working rhythm.

The intellectual demands to operate a machine should match the ability of the operator, neither under-demanding nor over-demanding.

7.10 Temperature and humidity

The control of temperature and humidity in the workplace is important in ensuring optimum output. The environment of a workplace needs to be maintained within certain temperature and humidity limits for worker comfort and efficiency. Machines that generate heat and emit steam to the extent of interfering with the workplace environment should be provided with measures to reduce that effect, such as ventilation or extraction plant.

7.11 Ventilation

Heat and fumes emitted by machines can have an adverse effect on output and may be a health risk to the operators. Extraction units should be provided to remove the fumes and provision made to supply clean fresh air to replace that extracted.

7.12 Repetitive actions

Frequent and continuous muscular movement of the hands, wrists and arms can result in a number of disorders known as *work related upper limb disorders* (WRULD). Machines should be designed so that repeated movements are not necessary or, if they do occur, do not require frequent muscular effort.

7.13 Warnings

Machines should carry permanent signs or notices warning of hazards associated with their operation (EN IEC 61310–1). The warning notices may be accompanied by a text and should be located in the region in which the hazard they identify arises. Care must be exercised in the number of signs and notices since too many may confuse rather than inform.

7.14 Vision

Vision is important in the safe operation of machines. Any factor that impairs it or makes observations difficult should be eliminated or reduced to an acceptable level. Factors to consider include vibrations, steam emissions, dust emissions, badly placed controls, badly placed instruments, lighting, etc.

7.15 Radiations

Machines can emit electromagnetic radiations that can be harmful to the operator. They may arise indirectly from the condition (temperature) of the machine or directly from a process. Radiations can extend from non-ionizing radiations in the long wave low radio frequencies range through infrared, light, ultraviolet to the ionizing radiations of gamma and x-rays.

High frequency radio waves (microwaves) cause damage by internal heating of flesh. Machines, such as plastic welding and induction heating machines, that operate within this frequency range should have effective screening around the whole area of the electrode from which the radiations are emittted.

Infrared radiations manifest themselves as heat and suitable shielding should be provided for the operator.

Lasers present a particular hazard by virtue of the extreme intensity of the beam. The main vulnerable organ is the eye. Machines and equipment employing lasers should have effective interlocking devices to ensure that the beams cannot be energized until all the special protective screens are in place (EN 12626, 31252 and 31253) (*Figure 5.3*).

Ultra violet radiations (UV) can cause skin cancer and cataracts. UV is used in ink drying and sterilization processes where the machines should be completely screened so that no radiations, direct or indirect (reflected), can escape into the workplace.

Ionizing radiations (x-rays, gamma rays, etc.) can cause irreparable damage to internal organs and the blood. Equipment using these radiations should comply with the strict international recommendations based on preventing all possible exposure (ICRP publication 75, General principles for the radiation protection of workers).

7.16 Indicators and instruments

The operator should be provided with sufficient information about the condition of the machinery to ensure its safe operation. This information can be derived from instruments and condition indicators.

Where only a general indication of the state of the equipment is required, *qualitative* indicators can be used which show an expected normal operating condition with an indication of permitted high or low variations outside the normal range, i.e. temperature, pressure, speed, oil level, etc. These are suitable for most applications where sufficient indication of condition can be given by a mark on an instrument dial or face (safe working pressure of steam boilers).

Where an operator needs to know precisely the condition of the machine or equipment, the indicators should give *quantitative* readings in the form of numerical values which then require the operator to interpret the importance of the information given. This type of indicator may require audible or visual warnings to draw the operator's attention to variations outside a normal permitted operating range. Important gauges and indicators should be mounted in the foreground of panels in a position that is easily readable by the operator.

7.16.1 Indicating instruments

Indicating instruments can be:

(i) *Rotary* in which either a needle or pointer moves round a circular scale or a marked disc rotates about a fixed pointer. The pointer should be no wider than a single scale marking and should not impinge on the letters or figures. If the position of the pointer for normal operation is shown on the dial, i.e. safe working pressure of a boiler, variations from norm can be seen easily and quickly. Round indicators can give a good spatial indication, i.e. contents of a storage tank.
(ii) *Linear* indicators where a column or line moves against a linear scale can give a good indication of degree within a range but are not particularly accurate, i.e. sight glass contents gauges. If the scale is marked at the normal condition, they give a good indication of variations from it.
(iii) *Digital* indicators are accurate and display figures in a small panel. They are quick and easy to read but do not give any indication of variation from norm.

7.16.2 Figures and letters

For clarity, figures and letters should be plain without crossbars, i.e. sanserif. For most applications black letters on a white background are easiest to read but in some reduced light situations white letters on a black ground show up better.

7.16.3 Indicator performance

In all cases, indicators should behave consistently and follow accepted practices, i.e. increases show as a clockwise movement on a round scale and movement to the right on a linear scale.

Where a number of gauges or instruments indicate conditions in different parts of the equipment, the position of each indicator, pointer or the position of the indicating line should all be the same for normal operating conditions so that any errant reading can be noted quickly and easily.

7.16.4 Indicator lights

Indicator lights can be used to show the existence of particular conditions, whether normal or abnormal. Flashing lights can be used to:

- attract attention;
- indicate action is needed;
- warn that the equipment is adjusting to new parameters;
- warn of discrepancies between control settings and actual conditions.

Operators must be trained in the implications of the different lights.

Indicator lights on instrument panels should not be so close together as to confuse and the function of each should be clearly identified. The intensity of each light should be sufficient to show up clearly under normal workplace lighting but not so bright as to dazzle or outshine other indicator lights.

Colour coding for indicator lights should follow IEC 60204–1; 10.3.2.

Colour	Meaning	Circumstance and action
Red	Emergency, stop	Danger, hazardous condition, operator to intervene to remove hazardous condition
Yellow	Warning, abnormal	Conditions vary from norm, operator to intervene and adjust controls
Green	Safe, normal	Machine and equipment operating within acceptable parameters
Blue	Mandatory	Operator must take action demanded by this indicator
White	General	No specific function but can indicate that equipment is live and operating. If it has a specific function operators must be trained in the action to take

7.17 Coda

In the design of a machine account must also be taken of ergonomic factors, which may be influenced by environmental factors such as noise, vibration, temperature, lighting and possible escape of hazardous materials from the process.

The physical layout of machines and their neighbours and the demands made on operators must be considered to achieve optimization of efficiency and output.

Reference

Kroemer, K H E., and Grandjean, E., *Fitting the Task to the Human*, 5th edn, Taylor & Francis, London, 1999.

Part III
Safeguarding systems

The function of the safety circuit is to react to the entry into, or the presence in, a danger zone of a person and initiate a signal to the machine controls that causes them to bring the machine to a safe state. This part deals with some typical safety circuits and the techniques employed covering the different media – mechanical, electrical, hydraulic and pneumatic – that can be used.

Chapter 8
Mechanical safety arrangements

8.1 Introduction

The object of a guard is to prevent a person or parts of a person from coming into contact with dangerous parts of machinery. This can be achieved by either providing a physical barrier, such as a guard or fence, between the person and the dangerous part or by other safeguarding arrangement that ensures that when a person reaches a part of the machinery it is no longer dangerous, i.e. it is not moving. This chapter deals with the design of mechanical guards and related devices. The designs of safeguarding arrangements for electrical, hydraulic and pneumatic circuits are dealt with in *Chapters 9, 10* and *11* respectively.

8.2 Guards

Guards provide a physical barrier between the operator and the dangerous parts of a machine. They can take the form of either enclosing guards or distance fencing.

8.2.1 Enclosing guards

Enclosing guards completely block all access to dangerous zones. They must be robustly constructed to act as an effective barrier and to withstand the operating environment for an economic period of time. Guards must:

- be of adeqate size to cover all means of access to hazardous parts;
- have suitable strength to retain their shape – depending on the size, shape and material used they may require supporting by a frame;
- be of a material that will resist wear and tear and not allow access;
- not interfere with the operation of the machine;
- not cause failure of the machine, i.e. from overheating of motors or belt drives due to lack of ventilation;
- if in the form of a fence, be of suitable height and positioned far enough away from the dangerous parts to ensure that these cannot be reached;
- have provisions to allow cleaning of the protected areas.

108 Safety With Machinery

8.2.2 Materials

Suitable materials include:

- sheet metal but this may not be suitable where ventilation is necessary to keep machine parts from overheating, i.e. vee belt drives (*section 3.3.3(d)*);
- weld mesh (*Figure 8.1*), has considerable strength and can be shaped to suit but mesh sizes must not allow the dangerous parts to be reached. It permits viewing of work inside the guard. Viewing through weld mesh is enhanced if the mesh is painted matt black and a light placed behind it. Chickenwire and woven wire are not suitable as fencing materials since they do not have the necessary strength and can easily be distorted;
- expanded or perforated metal of suitable gauge can be shaped but may require mounting on a frame to give support (*Figure 8.2*). It must be treated to remove burrs and sharp edges;
- polycarbonate and similar plastic materials with a high impact resistance tend to be soft and while they can be formed easily also scratch easily. Because they have good dielectric properties they also attract dust. They are useful where clear viewing through the guard is necessary as shown in *Figure 8.3*. Because they can be cleaned easily, they have particular application in the food and pharmaceutical industries.
- perspex is a hard brittle plastic which can shatter when struck and generally is less suitable than polycarbonate for guards;

Figure 8.1 Guard constructed of weld mesh. (Courtesy W Smith Packaging Ltd.)

Mechanical safety arrangements 109

Figure 8.2 Guard constructed of expanded metal. (Courtesy of the Allan Group.)

Figure 8.3 Polycarbonate guard showing the clarity of view of the equipment behind it. (Courtesy of Marstan Press Ltd.)

110   Safety With Machinery

Figure 8.4 Moulded guard for an electric lawn mower. (Courtesy of Black & Decker Ltd.)

- timber and hardboard either as panels or boards are easy to manipulate and can be made into rigid structures but may have a limited life. They are flammable and should not be used where there is a fire risk;
- shaped thermosetting plastics can be self-coloured and moulded to provide an attractive profile particularly to manually operated tools as shown in *Figure 8.4*.

Where the guard material has gaps and openings, the guard must be positioned such that it is not possible for any part of the body to reach any dangerous part. In this, allowance must be made for the variation in sizes of body parts that exists between different ethnic groups.

8.2.3 Gaps in guards

Where guards are designed either of material containing gaps or with gaps as an inherent part of the guard, the guard should be positioned so that it is impossible to reach through the gap and make contact with a dangerous part. In general, slots allow deeper penetration than round, square or rectangular openings. *Tables 8.1 and 8.2* summarize the safe distance at which openings of various sizes should be located to prevent hand, arms and legs respectively from reaching a dangerous part.

Mechanical safety arrangements 111

Table 8.1 Diagram of safety distances for hand and arm reach through a guard. (Table 4 from EN 294, courtesy British Standards Institution)

Openings in guards for 14 year olds and above

Dimensions in millimetres

Part of body	Illustration	Opening e	Safety distance sr		
			Slot	Square	Round
Finger tips		$e \leq 4$	≤ 2	≤ 2	≤ 2
		$4 < e \leq 6$	≥ 10	≥ 5	≥ 5
Finger to knuckle joint		$6 < e \leq 8$	≥ 20	≥ 15	≥ 5
		$8 < e \leq 10$	≥ 80	≥ 25	≥ 20
		$10 < e \leq 12$	≥ 100	≥ 80	≥ 80
		$12 < e \leq 20$	≥ 120	≥ 120	≥ 120
or		$20 < e \leq 30$	$\geq 850^1$	≥ 120	≥ 120
Hand					
Arm up to junction with shoulder		$30 < e \leq 40$	≥ 850	≥ 200	≥ 120
		$40 < e \leq 120$	≥ 850	≥ 850	≥ 850

[1] If the length of the slot opening is ≤ 65 mm the thumb will act as a stop and the safety distance can be reduced to 200 mm

Table 8.2 Diagram of safety distances for leg reach through a guard. (Table 1 from EN 811, courtesy British Standards Institution)

Part of lower limb	Illustration	Opening e	Safety distance sr in mm	
			Slot	Square or round
Toe tip		$e \leq 5$	0	0
Toe to ball of foot		$5 < e \leq 15$	≥ 10	0
		$15 < e \leq 35$	$\geq 80^1$	≥ 25
Foot to ankle		$35 < e \leq 60$	≥ 180	≥ 80
		$60 < e \leq 80$	$\geq 650^2$	≥ 180
Leg up to knee		$80 < e \leq 95$	$\geq 1100^3$	650^2
Leg up to crotch		$95 < e \leq 180$	$\geq 1100^3$	$\geq 1100^3$
		$180 < e \leq 240$	Not admissible	$\geq 1100^3$

[1] If the length of the slot opening is ≤ 75 mm the distance can be reduced to ≥ 50 mm.
[2] The value corresponds to 'Leg up to knee'.
[3] The value corresponds to 'Leg up to crotch'.

Mechanical safety arrangements 113

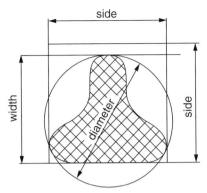

Figure 8.5 Nominal size rating for irregular openings. (Figure 1 from EN 811 courtesy British Standards Institution.)

The rating of sizes of openings of irregular shape should be assessed according to *Figure 8.5* by:

(a) determining
 - the diameter of the smallest round opening,
 - the side of the smallest square, and
 - the width of the narrowest slot
 that will encompass the irregular shape;
(b) match each with the safety distances s_r in *Tables 8.1* and *8.2*;
(c) use the shortest of the three safety distances identified.

If for any reason it is considered necessary, and provisions are made, to reach through a guard with a hand or arm, the design must ensure that dangerous zones cannot be reached. *Table 8.3* shows typical reaching configurations with associated safety distances.

8.3 Distance fencing

In some applications it may be simpler to surround sections of the machinery with a fence. The fence must be of a size and positioned to prevent dangerous parts being reached. Account must be taken of the ability of operators to reach over and under it. Any gap left below the bottom of the fence should not permit whole body access, i.e. be less than 186mm. Special arrangements may need to be made to allow access under the guard for housekeeping and cleaning purposes but they should not compromise the level of safety.

If access is necessary for setting, adjusting, maintenance or cleaning it should be provided by suitable interlocked gates. With high risk machinery additional protection, such as pressure mats (*section 6.12*) or photo-electric curtains (*section 6.10*), should be suitably positioned inside the fence.

Table 8.3 Safety distances when reaching through gaps in guards (Table 3 of EN 294 courtesy of British Standards Institution)

Limitation of movement	Safety distance sr	Illustration
Limitation of movement only at shoulder and armpit	≥850	
Arm supported up to an elbow	≥550	
Arm supported up to wrist	≥230	
Arm and hand supported up to knuckle joint	≥130	

A: range of movement of the arm.
[1] Either the diameter of a round opening, or the side of a square opening, or the width of a slot opening.

8.3.1 Fencing and other distance barriers

These must be of a size and position to prevent access to the danger zone. The height of the fence will determine the distance from the danger zone. *Table 8.4* gives safety distances for a range of heights of fences.

Table 8.4 Safety distances when reaching over barriers (Figure 2 and Table 2 from EN 294 courtesy of British Standards Institution.)

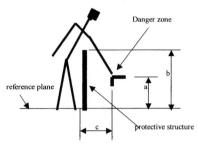

Height of danger zone	Height of protective structure b[1]					Dimensions in millimetres				
	1000[3]	1200[3]	1400[3]	1600	1800	2000	2200	2400	2500	2700
2700[2]	–	–	–	–	–	–	–	–	–	–
2600	900	800	700	600	600	500	400	300	100	–
2400	1100	1000	900	800	700	600	400	300	100	–
2200	1300	1200	1000	900	800	600	400	300	–	–
2000	1400	1300	1100	900	800	600	400	–	–	–
1800	1500	1400	1100	900	800	600	–	–	–	–
1600	1500	1400	1100	900	800	500	–	–	–	–
1400	1500	1400	1100	900	800	–	–	–	–	–
1200	1500	1400	1100	900	700	–	–	–	–	–
1000	1500	1400	1000	800	–	–	–	–	–	–
800	1500	1300	900	600	–	–	–	–	–	–
600	1400	1300	800	–	–	–	–	–	–	–
400	1400	1200	400	–	–	–	–	–	–	–
200	1200	900	–	–	–	–	–	–	–	–
0	1100	500	–	–	–	–	–	–	–	–

[1] Protective structures less than 1000 mm in height are not included because they do not sufficiently restrict movement of the body.
[2] For danger zones above 2700 mm other safety measures should be used.
[3] Protective structures lower than 1400 mm should not be used without additional safety measures.

Note 1: Fences below 1400 mm (56 in) should be avoided since there is a risk of toppling over them. If a fence of height less than 1400 mm is used, additional protective measures must be provided within the fence to prevent dangerous parts being reached.

Note 2: The gap between the bottom of the guard and the floor should be less than 186 mm and at least 650 mm from any dangerous part to prevent toe access.

Note 3: The minimum height of a safety fence must not be confused with the recommended height for a hand rail (1100 mm) which may be provided for guidance, support or as a deterrent barrier (*section 5.3.14*).

8.4 Safety gaps

Where it is necessary to interpose parts of the body between machine parts, certain minimum gaps must be maintained to prevent crushing of the body part *(Table 8.5)*. The minimum distance can be maintained either by

Table 8.5 Clearance required between parts of machinery to prevent body parts being crushed. (Table 1 from EN 349, courtesy of British Standards Institution)

Part of body	Minimum gap (a)	Illustration
Body	500	
Head (least favourable position)	300	
Leg	180	
Foot	120	
Toes	50	
Arm	120	
Hand Wrist Fist	100	
Finger	25	

inherent design such that when closing parts reach the end of their travel an adequate gap remains, or by the insertion of a mechanical scotch (*section 5.3.13*) of suitable length. Scotches should not be used when the parts are under power drive but only to hold them apart when power is removed, e.g. to hold the platen of a down-stroking press in the raised position to allow access for setting, adjustment or maintenance of the tools.

8.5 Feed and take-off stations

Automatic feed and take-off stations should be fully guarded by tunnel guards (*Figures 5.10* and *8.6*) or similar means that prevent access into the dangerous parts.

Manual feed stations must be provided with a suitable guard or safeguarding device that ensures that the operator cannot reach parts of the machine when they are still dangerous. Any gaps in physical guards must not exceed the dimensions given in *section 8.2.3*. Where the feed is by means of a power driven loading table that transfers the material to the working part of the machine, the design of the transfer mechanism should ensure there is no trapping hazard or if a trapping hazard is created it must be protected by a suitable trip device that stops the mechanism before injury occurs. Guarding at manual take-off must not allow the operator to reach into the dangerous operating part of the machine.

Similarly with automatic ejection of product it must not be possible to reach the dangerous parts with or without the receiving bin in position. Where components are ejected down a metal chute, the noise generated can be reduced by lining the chute with suitable sound deadening material.

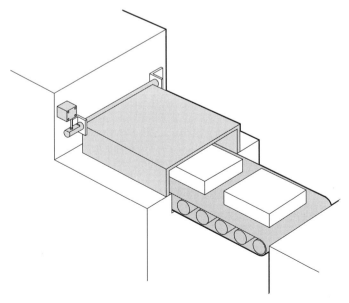

Figure 8.6 Tunnel guard at delivery from a machine. Note: the guard can be hinged and interlocked to allow access to clear jam-ups etc.

8.6 Work-holding devices

Manual clamping devices rely on the skill of the operator to ensure they remain secure. However power operated work-holding devices, whether manually controlled or automatic, rely for their security on the continuity of the power supply. They should be designed so that, on loss of power, the workpiece remains securely held, a warning is given and the machine is brought to a stop. There should be visual indication of the status of the power supply. On guillotines with manual gauging, i.e. the clamp is positioned under manual control, power clamping should only occur while the manual clamp is applying pressure. Any relaxing of manual gauging pressure should automatically prevent the power clamp from being activated.

With automatically fed machinery, the clamping pressure should be interlocked with the machine controls so that in the event of loss of clamping power the machine cannot start. Conversely, the clamp should not release until the machine is in a safe condition to allow it to do so. In both cases, an indication should be included on the control panel of the state of clamping pressure and that the the workpiece is correctly located. Power clamping and the processing operations should be fully guarded.

8.7 Counter-weights

Where counter weights are provided for a rise-and-fall guard, a moving component of a machine or to provide tensioning to a belt or rope, the weights should be arranged so that as they move they cannot trap or strike parts of the operator. Normally this can be achieved by containing the weights within a suitably shaped tube which covers the full extent of the movement of the weight and either be fitted with an end stop or continue to the floor or a suitable part of the machine frame (*Figure 3.3.2.d*).

Suspension of the counter-weights can be by wire ropes, chains or belts but care must be taken to ensure that any pulleys round which the supporting ropes, etc. pass are of a diameter and construction that is suitable for the suspending means. Where there is more than one suspension support each should be capable of taking the total weight and there must be means of adjustment so that loads can be equalized.

The counter-weights can be replaced by constant tension spring devices of a rating to support the weight or apply the required load (*Figure 5.3*).

8.8 Safety catches

Where it is necessary to retain a guard or part of a machine in an elevated position to allow work to be carried out below it, that guard or part must be retained by a secure device that cannot slip out of engagement. To release the catch, the item being held must be raised, i.e. its weight taken by the supporting rope, etc., before the latch can be withdrawn. A positive action must be required of the operator to release it. The device can be a gravity operated manual release device, a spring loaded ratchet or a power operated bolt or catch.

To prevent run back of either a sliding or rotary movement, a simple spring loaded ratchet with manual override can be used.

On mechanical presses where the ram or platen stops at top dead centre, run-back or over-run due to gravity can be prevented by the use of a deadlock device inserted below the upper platen or by a progressive scotch (*Figure 5.13*). In both cases the device must be robust enough to support the full weight of the ram plus tool.

8.9 Braking systems

To bring a machine to a stop quickly requires a brake of one sort or another. The type used will be dictated by the machine and its control system. Because a brake works by the absorption or dissipation of energy, the parts to be brought to a stop should have as low a momentum as possible. Since a brake may be part of an emergency system it should be capable of bringing a machine to a stop from the maximum speed of operation. It must be capable of dissipating any heat generated by friction and still retain its effectiveness. It is important that brakes perform consistently. Where that performance is critical to safety, stopping performance should be monitored (*section 9.8.5*) and interlocked with the machine controls. If allowed stopping time is exceeded further start up should be prevented. Care must be taken in the design to ensure that all the parts of the machine are capable of withstanding the reverse stresses resulting from the action of braking and that screwed fastenings cannot become undone.

8.9.1 Mechanical brake

Mechanical brakes work by the application of a friction material against a metal drum, rim or disc that is attached to the moving parts, usually the main drive shaft, of the machine. Braking pressure should be applied by two or more springs acting on the components to which the friction material is attached. Brake release by compressing the springs can be achieved by mechanical, electrical, hydraulic or pneumatic means, thus allowing the machine parts to move. Whichever system is used, it must be capable of releasing the springs instantly in an emergency. Failure of power supplies should cause the instant application of the brake.

The springs used should be closely matching sets and compressed to maker's instructions. Efficient locking means must be provided on the compressing bolts to prevent their slackening back. Written instructions should be provided detailing the procedure for setting the operating mechanism.

Alternatively, braking pressure can be applied by hydraulic or pneumatic means but this will require a special dedicated pressure storage vessel serving only the braking circuit with a non-return valve in the vessel pressure supply line. There should be a pressure switch in the vessel set so that if the vessel pressure falls below a pre-determined level it trips the machine. A pneumatic safety circuit containing a dedicated pressure storage vessel is shown in *Figure 11.6*.

In certain applications, such as the hoist drums or discs of cranes where the brakes are solenoid released, the control circuits must be arranged so that the solenoid does not release the brake until the hoisting drum has taken the weight of the load. Mechanical brakes have the advantage that they are able to retain the machine in its stopped position.

8.9.2 Electrodynamic braking

Electrodynamic braking systems utilize the existing electrical power circuits of the machine to bring the machine to a halt. They rely on a continuation of electrical supplies for their effective operation, any interruption of supply can result in failure of the braking device. Therefore these devices must be provided with a separate secure (dedicated) power supply. Actuation can be by operator control but in safety applications is by a trip device on the machine. A typical safety application is shown in *Figure 8.7* but this does require suitable control equipment (*Figure 6.22*). Proprietary braking systems use one or a combination of the following braking methods.

8.9.2(a) dc injection

Applicable to ac motors, on actuation the ac supply to the stator windings is disconnected and replaced by a dc supply which rapidly brings the rotor to rest. To provide the required degree of safety integrity a secure dc supply is necessary.

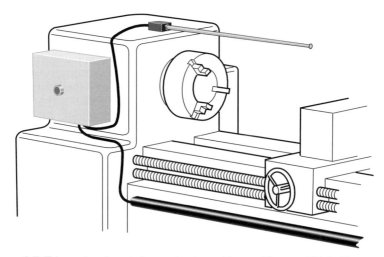

Figure 8.7 Telescopic trip switch on a horizontal borer. (Coutesy EJA Ltd.)

8.9.2(b) Reverse plugging

In this method the electrical connections to the motor are switched to a configuration that would cause the motor to run in the reverse direction. This causes the motor to come to a rapid stop but if the reverse switching is left on there is a danger that the machine will start running backwards. A motion sensor or other device is necessary to open circuit the braking power supply contacts as soon as all motion has stopped.

8.9.2(c) Regenerative and dynamic braking

In principle, regenerative braking dissipates the electrical energy that can be generated when a motor over runs, i.e. is driven by the machine it was driving. This regenerated power can be fed back into the mains. Variable speed electronically controlled drive motors – particularly crane drives – are braked by using the electronic controller in inverse control mode so that the regenerated power is directly fed back through the controller into the mains.

8.9.2(c)(i)

For ac motors, when braking is required, the motor terminals are connected to a capacitor bank. This maintains the stator excitation when the mains are removed but as the speed of the motor drops so the effectiveness of the braking reduces. When the motor reaches a predetermined speed, the motor terminals are short circuited and after a further predetermined time dc injection is applied to bring it to a rapid stop.

8.9.2(c)(ii)

With dc motors, the motor terminal connections can be switched so that the motor acts as a generator. The power generated can be fed back into the main power supply or connected to a dump load (dynamic braking resistor).

8.10 Clutches

Clutches are the mechanical means for connecting and disconnecting a power source to or from the machine it drives. They can be operated manually, mechanically, electromagnetically, hydraulically or pneumatically and are normally of the friction plate type. However, other types such as dog clutches, magnetic drive clutches and fluid driven clutches (fluid couplings) are also used. Whatever the type, the clutch must be capable of meeting the duty required of it without excessive wear or fatigue, be regularly maintained and kept free from substances that would cause its performance to deteriorate.

The main two types of duty performed by clutches are the continuous transmission of power and the cyclic transmission of power for a single stroke or short cycle operation. In single stroke press brakes it is usual for the clutch and brake to be in one common unit.

In continuous running machines, the clutch is normally engaged whereas in single stroke machines the clutch is normally disengaged and only engages when the machine is struck-on.

On positive clutch power presses where the clutch is one of the elements of a safety system it should be linked to the guard or safeguard device in such a manner that the clutch cannot be engaged until the guard is closed, i.e. in the safe operating position. Once the operating cycle has commenced it should be impossible to open the guard until the cycle is complete and the ram has stopped at top dead centre. A typical guarding arrangement for a positive clutch press is shown in *Figure 8.8*.

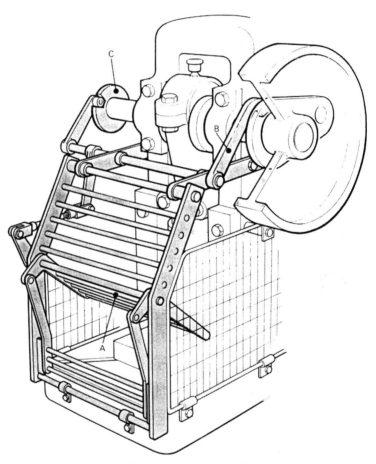

Figure 8.8 Interlocking guard for a positive clutch power press. The guard consists of an enclosure with a movable gate A. When the gate is closed the guard prevents access of any part of the body to the danger area from any direction. The gate is interlocked by lever B with the clutch mechanism in such a way that the press cannot operate until the gate is fully closed. While a stroke is being made the gate is held closed by the guard control C and cannot be opened until the clutch has disengaged and the crankshaft has come to rest at the correct stopping position, usually at top dead centre. (Figure 99 from BS PD 5304.2000 courtesy British Standards Institution.)

8.10.1 Dog clutches

Dog clutches work by a positive mechanical engagement of two interfitting parts. Engagement can only be made when the two parts are stationary or rotating at the same speed – attempts to engage when one part is moving and the other not, can result in excessive shock stresses with the likelihood of failure of machine parts. Where manually operated hand wheels with dog clutches are provided on a running shaft of a machine for purposes of movement during setting and adjustment (*Figure 3.3.1(i)*), the clutch should be spring loaded out of engagement and engagement not attempted until the machine has been brought to a stop.

8.10.2 Magnetic drive

In this arrangement, magnetic forces, often electrically induced, are the means by which power is transferred from one part of a machine to another part. Care must be taken to ensure the magnetic forces do not interfere with the controls of other machines, adjacent electronic equipment or with heart pace makers.

8.10.3 Hydraulic drives

Hydraulic drives can be of two basic types – fluid coupling and hydraulic motors. *Fluid couplings* employ the shear of a fluid between radial vanes on adjacent discs to transmit power and are used where a slow take up of power from a relatively fast rotating power shaft is required. In an emergency they can be disconnected quickly but a brake is necessary to bring the driven machine to a stop and hold it stationary. Fluid levels must be maintained for optimal effectiveness.

Hydraulic motors operate through positive displacement hydraulic pistons driving a shaft. Very high levels of torque at low speeds can be transmitted with a very fine degree of control. Shutting the control (feed) valve stops the motor and hydraulically locks it in the stationary condition.

8.10.4 Plate clutches

The most common type of clutch utilizes a friction material between two metal plates to achieve transmission of power. To give optimum performance, the friction material must either be kept free from substances that could affect its friction properties by providing suitable protective covers, or the friction material must be selected for its resistance to the contaminating substance. Normally the clutch plates are spring loaded together and disengagement is by separating the plates using one of a number of media. In special applications, such as press brakes, the clutch and brake may be combined in one unit with the brake in normal engagement. Brake release and clutch engagement only occurring when the machine is struck on.

8.10.4(a) Manual control

Actuation is by means of a mechanical link from the control position to the clutch and allows for progressive application. This method is suitable for drives where selective control is required to match changing operating conditions but is not suitable as part of a safety system.

8.10.4(b) Mechanical control

To use in a safety application, the clutch actuator must be directly linked to the machine (press) guard such that the press will not strike on until the guard is closed and, once struck on, the guard cannot be opened until the machine has completed its full operating cycle. In the guard open condition, the clutch actuator should be mechanically restrained in the clutch disengaged position (*Figure 8.8*).

8.10.4(c) Electromagnetic control

Actuation of the clutch is by energizing an electromagnetic solenoid. Any reduction or loss of power to the solenoid should cause the machine to stop. The solenoid circuit should be linked with the guard and safety circuits. The clutch and brake may be operated by a common solenoid.

8.10.4(d) Hydraulic control

As an incompressible fluid the connection from the control to the actuator is 'hard' and effectively is similar to a mechanical link with the advantage of flexibility in the routing of the control pipe. In a safety application where the clutch is held in engagement by the hydraulic pressure, any loss of that pressure should cause the machine to stop. Provision must be made in the design of the hydraulic circuit to accommodate emergency stopping.

8.10.4(e) Pneumatic control

Operates in a similar manner to hydraulic control but has the disadvantage that, as air is a compressible medium, movement of the actuator can be irregular and responsive to the condition of the operating cylinder.

8.11 Summary

Mechanical safety arrangements are an effective method for ensuring the safety of those who operate machines, but their continuing effectiveness relies on the devices being kept in prime working condition. High standards of maintenance are essential and can best be achieved by regular planned maintenance of the safety-related and safety-critical parts of the machines and, indeed, of the whole machine.

Chapter 9
Electrical safety circuits

9.1 Introduction

The control system of a machine is made up of those components – mechanical, electrical, hydraulic or pneumatic – that, when actuated, cause the working parts of the machine to function [IEC 16508 - 4 cl. 3.5 and IEC 60204 – 1 cl. 3.8]. A control system contains the following elements:

- a power source;
- control devices to transmit a signal giving an instruction or intention to
- the control component or circuit that processes the signal and causes
- the final power control element(s) to initiate machine action.

These elements are shown diagrammatically in *Figure 9.1*.

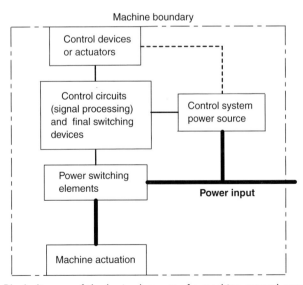

Figure 9.1 Block diagram of the basic elements of a machine control system.

125

9.2 Effect on safety

Those parts of the control system that are *safety critical* or *safety related* will need to be identified so that particular attention can be given to their design. Such parts are crucial to the safe operation of machines and for the protection of the operators during setting, adjustment and maintenance when the normal safety protection systems may be ineffective or isolated.

Identification of the crucial parts of the control system can most effectively be achieved by following the analytical procedures described in *Chapter 4*. The 'WHAT-IF' technique should allow a conclusion to be arrived at as to which parts of the control system are *safety critical or safety related*.

A more rigorous analytical technique will be required to establish the overall safety quality of the control system. With more complex forms of control systems, such as those which include electronic components, a substantial engineering resource will be needed to carry out this analysis adequately. Normally, the manufacturer of the control equipment would do this work and specify any safety limitations for its use.

9.3 Basic safety requirements

Switches with contacts, often referred to as 'limit switches', used as part of safeguarding systems and emergency stop switches must be of the positive disconnection type (*Figure 6.4*), or, if they are of the *fast break* type, must include a positive disconnection element. They must not fail to danger, i.e. failure of any part must not result in the safety circuit contacts remaining closed when the switch is actuated. In particular, failure of the return spring should cause the contacts to drop out. If the contacts are welded together the mechanical force from the actuated plunger should be sufficiently powerful to either part the weld or break the connecting link. Proximity (contactless) switches, microswitches or other forms of interlocking device must be of the *safety application* type.

On machines with more than one control position, the control circuit should be arranged so that, at any one time, only one control position is capable of starting the machine. This could be by selection at the main control position or by local selection which then isolates all the other control positions. In addition, the machine should be provided with an audible warning of impending start up and a delayed start function regardless of which position initiates the start (*section 6.20*).

Foot operated switches, whether for single stroke cycle or continuous run, should be arranged so they cannot be inadvertently operated. This can be achieved by the use of suitably designed shrouds or covers with an access from one direction only. Where the foot control is on a flexible lead the shroud design should be of a size that it will accommodate one foot only. The shroud or cover must be of robust construction capable of withstanding the operator's weight since collapse of the shroud could result in actuation of the machine. On some machines, such as power operated guillotines where the gauging pedal for clamping the work is foot operated, it should be of a design that allows its easy actuation from any position across the whole width of the machine

Where the control actuator is a handle, lever, rotary knob or handwheel, its movement must be consistent with the resulting effect (7.3) and it must be sufficiently clear of other similar controls to prevent their inadvertent operation.

9.4 Selection of interlocking switches

Interlocking switches are those components of a safeguarding system that are actuated by guard movement and connect with the machine controls. In the safe condition their contacts should be closed.

Typical interlocking devices used in conjuction with guards include:

- roller or plunger operated switches (and valves) on moving guards (*Figure 5.6*);
- proximity switches (*section 6.17*);
- tongue operated guard switches (*Figure 6.5*);
- hand operated delay bolts (*Figure 6.6*);
- hand operated key locked switches (*section 6.19*);
- key exchange switches (*section 6.18*);
- captive key switches (*Figure 6.24*);
- plug and socket links on guards (*Figure 9.2*);
- guard locking device (*section 6.3*);
- telescopic switch (*section 6.16*).

Fulfilling a similar function, but not involving guards, are trip devices:

- emergency stop switch (*section 6.15*);
- foot operated stop switches (and valves);
- grab wire switches (*section 6.14*);
- pressure sensitive mats (*section 6.12*) and edges (*section 6.13*);
- person sensing devices and switches, both magnetic and electronic (*section 6.10*).

The operating medium of these devices can be electrical, pneumatic or hydraulic or a combination.

In the plug and socket device multi-pin plugs should be used with the plug and socket pins interconnected to form a coding which cannot easily be defeated by short circuiting the pins. An alternative method is to use a diode connected in the plug and a control circuit from the socket that will not switch when a short circuit is applied to the plug (*Figure 9.2*). When used with a sliding guard, the guard should be arranged to cover the socket when moved to the open position.

All devices should be of a design that is suitable for safety applications and have been manufactured to a known safety standard. Devices without this assurance should be monitored to ensure that an acceptable level of safety performance is maintained.

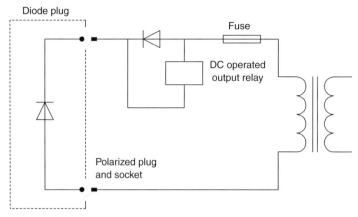

Figure 9.2 Circuit diagram for a diode plug interlocking device.

Where the actuating mechanism is external to the safety device it must be installed to meet the same level of safety performance as the device itself. Switches and their actuators should be securely fixed so they cannot become loose accidentally. Guard safety devices should not be easy to defeat. However, during maintenance and repair, skilled maintenance staff may be permitted to override safety devices but must take other, equally effective, precautions.

Switches must be installed so that the limits of actuator travel are within those specified by the switch manufacturer. Movement of external actuator travel, whether cam, roller or push rod should be limited to prevent damage to the switch mechanism.

9.5 Switching contact requirements

For a machine to be run with safety, the safety switch contacts forming the safety circuit must be closed and the safety circuit completely made. If any of the safety switches become open circuited, the safety circuit will be broken and the machine must revert to a safe condition. In the minority of circumstances where switching devices have to be open on safety demand, the safety circuits must include additional monitoring to reveal failures of those switches.

Switching contacts must be suitable for the loading of the circuit in which they are connected. This applies to both *slow break* and *fast break* contacts. The loads may be resistive or inductive and the manufacturer of the switching device may need to be consulted about its suitability for a particular application. Where, for inductive loads, arc suppression devices are required they must not be connected across safety switching contacts.

Switching devices with positive guided contacts should be used in safety circuits to prevent cross connection of change-over contacts if this could lead to danger.

9.6 Factors influencing the selection of interlocks

Table 9.1 gives examples of some of the problems met in selecting interlocking devices and possible solutions.

Table 9.1 Examples of typical problems met in applications and possible solutions

Circumstances	Typical solution
Access needed: • frequently • occasionally	Control interlocking using cam or tongue operated switches Key exchange system
Uncertain guard alignment	Use locating pins and proximity switches
Guard is removable or becomes loose	Use two switches operating in opposite modes
Food hygiene requirements	Use waterproof switches to an appropriate IP rating
Machine has long run-down time	Apply guard locking with release by either rotation sensors or time delay device
Machine needs to be run with guards open for setting and adjustment	Use: • limited movement control or • hold-to-run control

9.7 Circuit fault protection

The items described in this section relate to possible single fault conditions that could lead to failure of the safety system.

9.7.1 Control system power source

The electrical equipment of machines, including control systems, should conform to the standard IEC 60204-1. Power supply for the safety related parts of the control system must not develop faults that could result in a risk of injury.

Where, as a matter of expediency to overcome breakdown problems, control circuits are modified, regular inspections of the safety-related circuits and components should be made to ensure the modifications maintain the required level of safety and that the modifications do not become permanent. Faulty components should be replaced as soon as possible.

9.7.2 Circuit protection

The electrical protection of the circuits must ensure that the circuit components are not stressed beyond the over-current conditions specified by the component manufacturer. This will reduce the probability of switching contacts welding and solid state switches failing to a short circuit condition.

9.7.3 Insulation failure

Circuit design must take account of the possible effects of insulation failure leading to short circuiting of safety related switching devices. These are sometimes referred to as *cross-connection faults* (*Figure 9.3*). Factors to take account of include:

- mechanical damage of insulating materials in both the cables and component assemblies;
- breakdown of insulation due to age, hostile environment or overheating causing tracking on or in the insulating material;
- insulation breakdown due to the incorrect specification of creepage and clearance distances between connecting terminals, switching contacts or contacts and other conductors including an earthed conductor.

The control supply voltage should be as low as is practicable to reduce the risk of insulation failure.

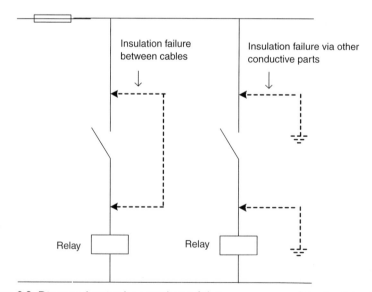

Figure 9.3 Diagram showing how insulation failure can cause a safety circuit to short circuit.

Insulation tests should be carried out regularly. To prevent damage to the equipment during insulation testing the procedures laid down by the manufacturers or suppliers should be followed.

9.7.4 Earth fault protection

Control circuits must be designed so that multiple earth (or ground) faults from short circuiting the switching elements of safety devices do not occur. This can be achieved by connecting one of the control circuit supply conductors to earth at the source of supply which ensures the automatic detection of the first dangerous fault.

Earth fault protection is inherent where a control circuit is supplied directly from an earth referenced mains supply. Where the control circuit is supplied through a control supply transformer earth reference is achieved by connecting the common connecting line of the control relay coils to earth at the transformer (*Figure 9.4*).

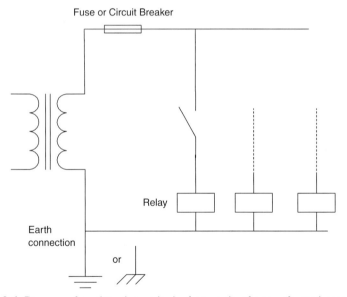

Figure 9.4 Diagram of earthing (grounding) of one pole of a transformed control supply.

The control circuit should be protected by either:

- a double pole circuit breaker to ensure that both poles of the supply are disconnected when the breaker trips, or
- a fuse or single pole circuit breaker in the non-earthed pole of the supply. The earthed conductor must not contain a protective device such as a fuse or single pole circuit breaker (*Figure 9.4*).

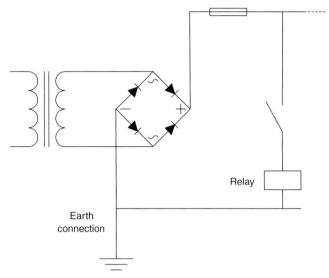

Figure 9.5 Diagram of earthing of one pole of a dc control supply.

With dc control circuits, where the supply is from a transformer through a rectifier, in order to prevent relays closing when an earth fault occurs, the earth reference must be connected to one of the dc conductors and not at the transformer (*Figure 9.5*).

If earth referencing of the control supply is not practicable, an equally effective alternative method must be used to maintain safety integrity. A circuit with double pole switching and over-current protection in each control supply conductor would provide an equivalent degree of safety (*Figure 9.6*). It requires two earth faults before the circuit protection

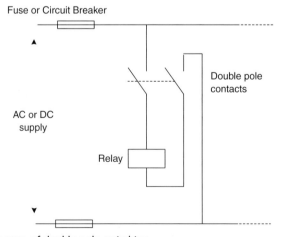

Figure 9.6 Diagram of double pole switching.

operates. If control circuit automatic disconnection is to be avoided, circuits must be tested regularly for earth faults.

Earth fault detection with automatic power supply disconnection or earth fault indication can be used (*Figure 9.7*). The earth leakage tripping current must be less than the current required to cause a safety relay, contactor or other actuator to stay closed if the earth fault(s) occur while they are closed. Where an earth reference connection has to be made at the mid-point of the control system power supply, protection against dangerous failures that can occur by centre-tapped

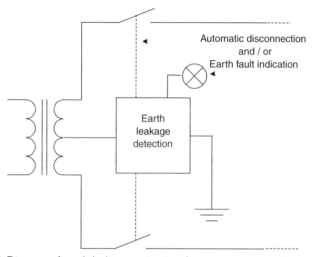

Figure 9.7 Diagram of earth leakage protection (overcurrent protection not shown).

direct earthing must be included. This can be achieved by making the connection to earth through a reliable high impedance earth detection circuit. The design and inspection methods must ensure that the first earth fault on the control circuit is detected and the machine is switched off or the fault repaired so that the machine remains in a safe condition.

Earth reference connections and earth monitoring equipment must be tested regularly to ensure safety integrity is maintained

9.8 Safety control circuits

The following descriptions provide an explanation of the various systems. The principles described can be applied separately to any of the three media of control or to a mixed media system. Block diagrams are used to illustrate the principles and some of the fundamental techniques are translated into circuit form. The circuits do not show overcurrent protection or voltage suppression devices and would need to be modified to suit a particular application.

The examples are based on a relay logic principle but safety control circuits, particularly if monitored, can be electronic logic systems providing they meet the appropriate level of safety integrity.

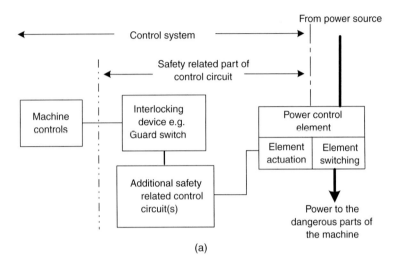

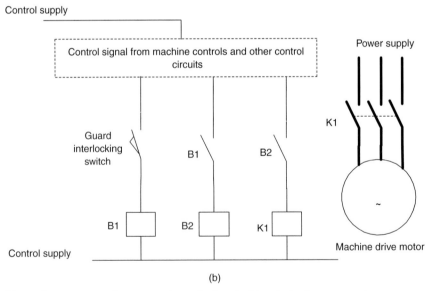

Note: there are some instances where the interlocking device is connected at a position in the circuit not close to the power control element. Failure modes of the interposing devices will influence the safety integrity of this type of crcuit.

Figure 9.8 Single channel control interlocking with additional safety related control circuits: (a) block diagram; (b) circuit diagram.

9.8.1 Single channel circuits

A single channel safety circuit consists of a single actuating device and control circuit in series and is suitable for application to a low risk machine. When actuated it either:

- causes the final power switching element to go to the open circuit condition, known as *control interlocking* (Figures 9.8 and 9.9), or

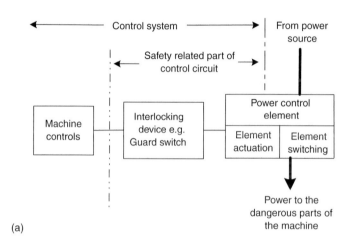

(a)

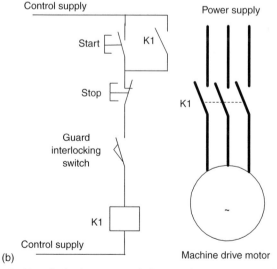

(b)

Note: The interlocking device is connected close to the power control element K1 which minimizes the number of failure modes that could lead to danger. It is the preferred method of connection for control interlocking.

Figure 9.9 Diagram of single channel control interlocking without additional safety related control circuit: (a) block diagram; (b) circuit diagram.

- is the final switching element in the power supply, known as *power interlocking* (Figure 9.10)

Regular maintenance is necessary to ensure that a failure of the safety system has not occurred or is not imminent and should include:

- the mechanical operation of the interlock actuator;
- the functioning of the interlocking device itself;

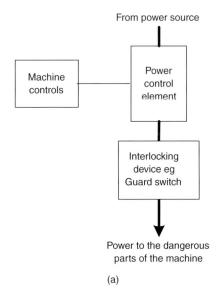

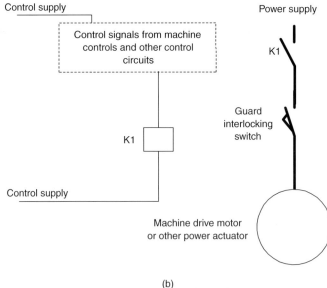

Figure 9.10 Single channel power interlocking: (a) block diagram; (b) circuit diagram.

- the condition of the final switching device, power switching element or power control element.

9.8.2 Monitored single channel circuits

To maintain or improve the level of safety performance in single channel safety circuits, they should be monitored by:

- passive monitoring using indicator lamps or other forms of alarm to warn the operator or supervisor of a malfunction (*Figure 9.11*);

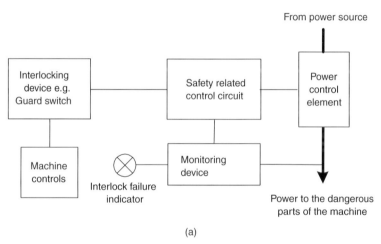

(a)

Figure 9.11 Single channel control interlocking with passive and active monitoring: (a) block diagram;

- automatic (active) monitoring whereby an automatic check of the safety circuit is made. The check should be carried out regularly and frequently, such as at the end of each operating cycle, whenever the machine is stopped and re-started or at the start of a shift, depending on the level of risk. The higher the risk the more frequent the check.
- If the control and monitoring systems are based on electronic principles, the check can be carried out while the machine is operating providing that the checking time – when the safety system will be ineffective – is short enough not to give rise to danger to the operator.

If the safety check fails, the machine should be stopped or brought to a safe condition and prevented from restarting until the fault is corrected (*Figure 9.11*).

138 Safety With Machinery

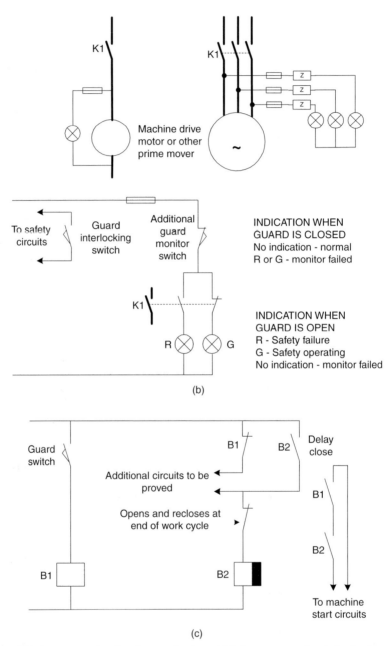

Note: This is an example of active monitoring which forces a guard switch check at the end of a work cycle or period. It would be of practical use where the failure of the monitoring relay B2 is not significant. If this relay has to be monitored the circuit becomes complicated and it would be more appropriate to consider the use of redundant circuit techniques.

Figure 9.11 (b) circuit diagram, passive safety monitoring using lamp indicators; (c) circuit diagram, active monitoring.

9.8.3 Dual channel circuits

Single channel safety circuits can give rise to the risk of injury if they fail during a machine's operating cycle. This risk can be reduced by introducing a second parallel but independent circuit – referred to as *redundancy* – since the probability of failure occurring in both circuits at the same time is greatly reduced. The safety integrity level can be increased by utilizing different control media, for example, one electrical and the other hydraulic. Passive monitoring is still required to warn of the first fault. Dual control interlocking is shown in *Figure 9.12* and dual power interlocking in *Figure 9.13*.

An alternative method to supplement the single channel circuit is automatically to divert the motive energy at the output side of the power switching element (*Figure 9.14* for electronic systems) and connect it to a low impedance load. This will prevent dangerous motion if the control circuit develops a fault. In mechanical systems the working fluids can be diverted back to the sump (hydraulics) or to atmosphere (pneumatics).

Regular inspections are still needed to maintain the level of protection. These dual channel safety circuits are *safety related* until one circuit fails when the other becomes *safety critical*.

9.8.4 Automatic monitored dual channel circuits

A higher level of safety integrity can be achieved by active automatic monitoring and *cross-monitoring* of the dual circuits (*Figure 9.15*). When a fault is identified in one of the circuits it causes the machine to go to a safe condition immediately or stop at the end of the current operating cycle.

A common way of achieving this is by using 'safety relays' which are manufactured by many companies who supply safety equipment. There are a range of products designed for particular safety applications.

9.8.5 Stopping performance monitors

Machines that are guarded by safety devices other than interlocked guards and which rely on braking systems to stop all motion after a safety device has been actuated, should be fitted with a *stopping performance monitor* to detect deterioration in the braking performance (*Figure 9.16*). When the stopping performance drops below a predetermined level the machine should automatically revert to a safe condition by either stopping immediately or at the end of the current operating cycle.

140   Safety With Machinery

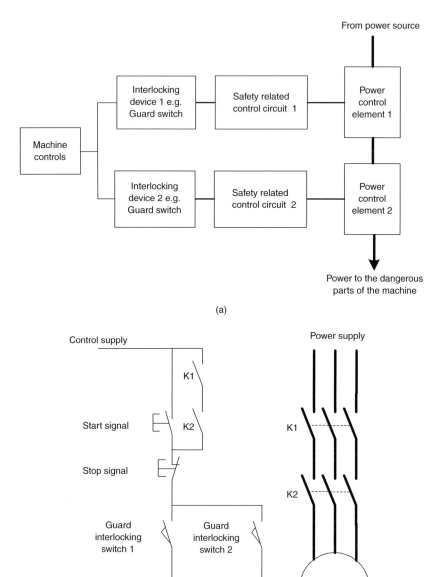

Note: The interlocking devices are connected close to the power control elements K1 and K2. In circuits where the interlocking device is not close to the power control element the common cause failure modes of the interposing devices will influence the overall safety integrity of the circuit.

Figure 9.12 Dual channel control interlocking. (a) block diagram; (b) circuit diagram.

Electrical safety circuits 141

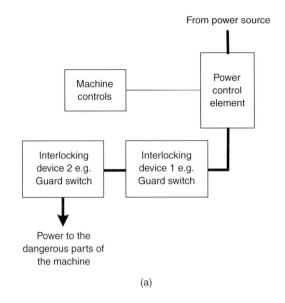

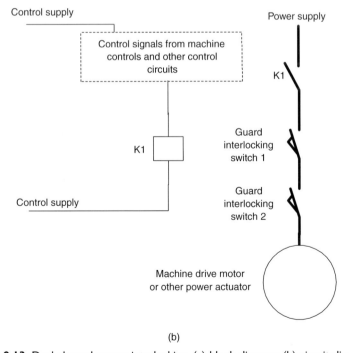

Figure 9.13 Dual channel power interlocking. (a) block diagram; (b) circuit diagram.

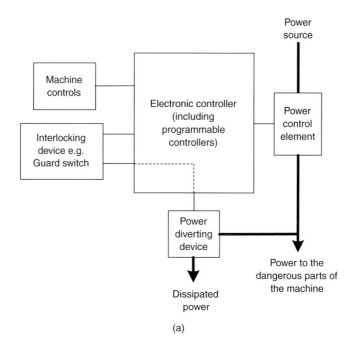

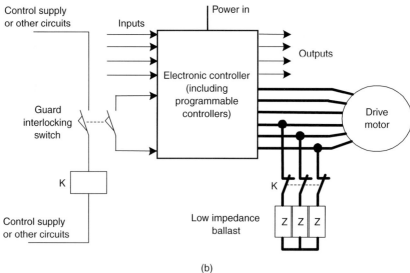

Note: The ballast and its switching arrangement has to be compatible with the type of drive control and output load.

Figure 9.14 Single channel interlocking with supplementary protection by power diversion: (a) block diagram; (b) example circuit diagram.

Electrical safety circuits   143

Figure 9.15 Dual channel control interlocking with cross-monitoring.

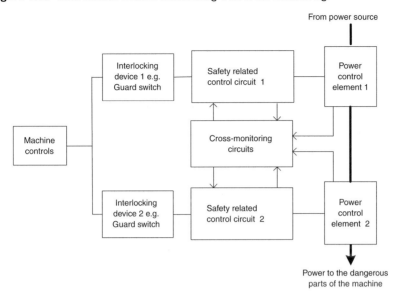

Figure 9.15a Block diagram.

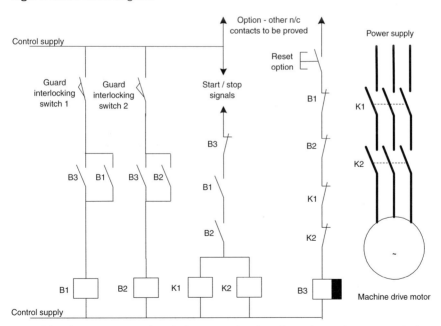

Note: B3 is the monitoring relay which ensures that the other relays and contactors in the circuit and any devices required to be monitored external to the circuit are in the 'off' state before the machine can be started at the beginning of a work period or after the guard has been opened and closed. There are other ways of configuring this circuit.

Figure 9.15b Circuit diagram incorporating an option for reset after the guard is opened.

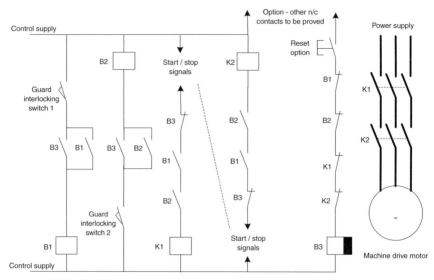

Figure 9.15c Modified circuit to prevent the dangerous effects of short circuits in the external wiring between the guard switches and the control system. Double pole switching is required for the start/stop signals.

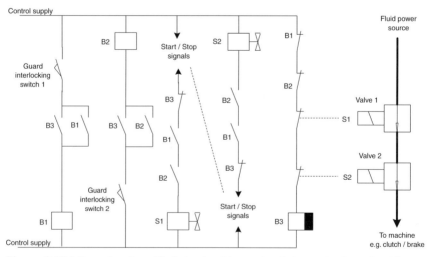

Figure 9.15d Example of modified circuit with guard switches and valve solenoids connected to reveal automatically the dangerous effects of short circuits in the external wiring between the switches, valves and the control system.

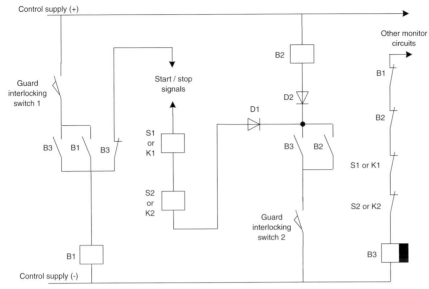

Figure 9.15e Alternative method of connecting the valve solenoids or power contactors.

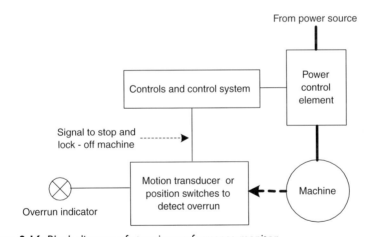

Figure 9.16 Block diagram of stopping performance monitor.

9.8.6 Delayed guard opening

Machines that are fitted with interlocking guards but which take time to run down to a safe condition when tripped should be provided with either:

- a guard restraint device that prevents the guard from being opened or moved until all the dangerous residual energy is dissipated or all motion has stopped; or

146   Safety With Machinery

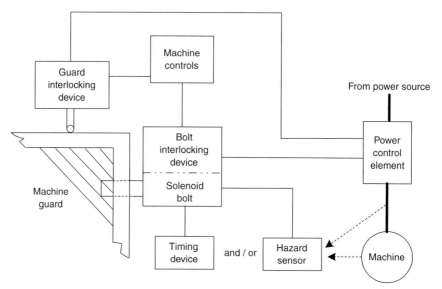

Figure 9.17 Block diagram of guard locking with interlocking switch and delayed release of lock until all residual energy has been dissipated.

- a time delay device that prevents the guard from opening until the expiry of a predetermined time after which all residual energy will have been dissipated or all motion stopped (*Figures 6.7* and *9.17*); or
- a manually operated mechanical time delay device (*Figure 6.6*).

Monitors and time delay devices are part of the safety system of the machine and should be designed so that component failure does not compromise safety. Their safety integrity should be to the same level as that of other safety devices and circuits on the machine.

9.8.7 Key exchange interlocking

Guard locking can be achieved through the use of a key exchange system in which the guard lock can only be opened by a key that is trapped in the isolating switch key box and cannot be removed until the isolator switch has been open circuited (*Figure 9.18*). Timing or motion detection devices can be incorporated in the isolator switch key board for machines with a long run-down time. Where there are a number of locked guards on a machine a master key box, which retains a number of keys, can be used (*6.18*).

9.8.8 Electronic safety control systems

Electronic safety control systems include programmable equipment, assemblies of electronic devices and components, *field bus* and cableless controls. The hardware and software of these systems can be complex

Electrical safety circuits 147

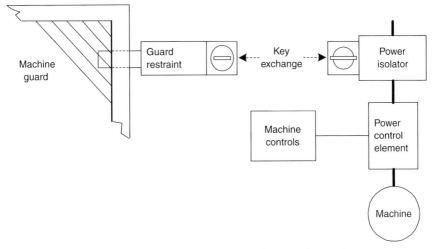

Figure 9.18 Block diagram of key exchange system for a single guard.

compared with the hard wired systems using *relay logic*. As a result there are many more possible failure modes so the system has to be designed to minimize the dangers that can arise from system faults.

All electronic equipment must be immune to the effects of electromagnetic interference, mains borne power supply interference and variations in power supplies that could have a detrimental effect on safety. The re-instatement of power supplies after a loss must not cause dangerous parts of the machine to start up.

Machine electrical systems should not emit radiation or create mains disturbance that could interfere with the operation of safety equipment of it or other machinery.

Safety software should be designed to minimize the probability of it failing to danger due to systematic errors (IEC 61508 – 3 and pr IEC 62061 cl.6).

Controller programs must be secured so that unauthorized operators cannot make changes that could lead to the risk of injury. As a general rule, all electronic safety control systems should, as a minimum requirement, meet the same level of safety as that achieved by the equivalent hard wired *relay logic* system.

In single channel circuits the interlocking device should be connected between the controller and the power switching device (*Figure 9.19*).

If it is not practicable to disconnect the power switching element without sending the same switching information to the electronic controller, a second contact of the interlocking device can be used to send the signal to the input of the controller (*Figure 9.19*).

Electronic controllers can be used as one or more of the safety-related channels of control systems but when used as a single channel they become safety critical and should incorporate automatic monitoring. The automatic discontinuous monitoring techniques described in *section 9.2.2* will satisfy the safety criteria for low risk applications.

Figure 9.19 Interlocking with electronic controllers.

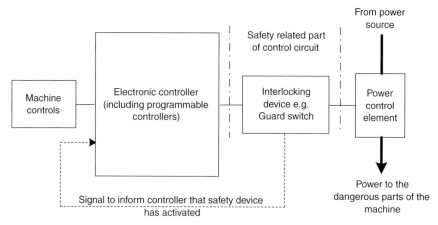

Figure 9.19a Block diagram.

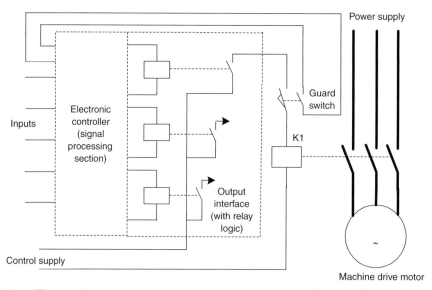

Note: The guard switches have an auxiliary switching element which provides a stop signal to the controller. An option is to use two guard switches which increases the safety integrity level.

Figure 9.19b Circuit diagram with the interlocking device in the output signal from the controller.

Electrical safety circuits   149

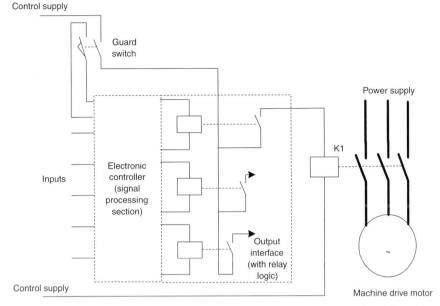

Figure 9.19c Circuit diagram with the interlocking device in the control supply to the output interface.

In multi-channel systems, diversity between channels – for both the software of programmable equipment as well as the electronic hardware – should be employed to reduce the probability of common mode failures.

Servo-controlled motor drives are complex control systems. Controllers for higher risk machine applications, such as robots, CNC machines, machining centres, etc., require continuous monitoring. Although these machines are provided with safeguards to prevent operators from approaching dangerous parts during normal operations, occasions arise, such as during machine setting and teaching, when moving dangerous parts have to be approached. In these circumstances, operators should use portable control stations fitted with hold-to-run controls and an emergency stop switch.

Where a remote diagnostic facility is included for either local or distant use, the safety of the machines must be maintained under all modes of diagnostic operations. The diagnostic circuits must be included in the safety assessments of the safety related machine control systems.

9.8.9 Emergency stop circuits

Emergency stop circuits should be manufactured to comply with IEC 60204-1.

All machines should be fitted with one or more emergency stopping devices (6.15) for use in the event of an emergency, unless it can be shown that such a device would not contribute to minimizing the risk. The

150　Safety With Machinery

Figure 9.20 Emergency stop circuits that are independent of the functional control system.

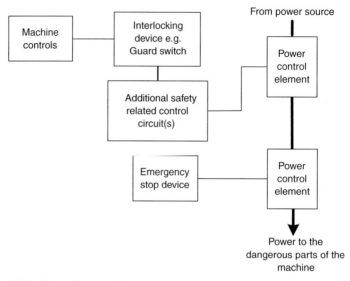

Figure 9.20a Block diagram of circuit that is independent of the functional control system.

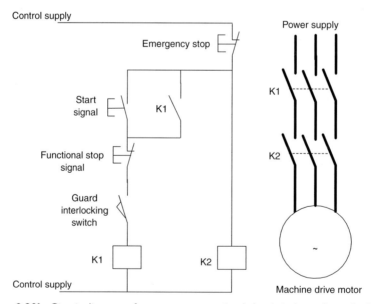

Figure 9.20b Circuit diagram of emergency stop signal that is independent of a basic functional control circuit and utilizes an additional emergency stop contactor.

Electrical safety circuits 151

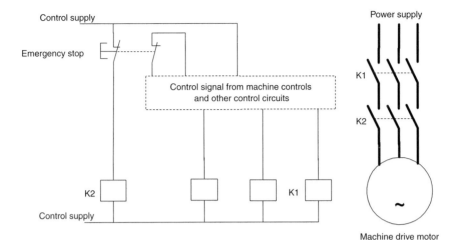

Figure 9.20c Circuit diagram of an emergency stop signal that is independent of complex functional control circuits and utilizes an additional emergency stop contactor.

number of channels that should be used for emergency stop circuits should be based on the findings of a risk assessment and on the resources available to inspect and test the circuits.

If an emergency could arise because of the failure of the machine functional safety control circuits to remove a risk, the emergency stop circuit(s) should be independent of the functional control system so that they interrupt directly the control signal to the power swtching elements causing it to go to open circuit (*Figure 9.20*).

Figure 9.21 Emergency stop circuit where the signal is processed by the functional control system

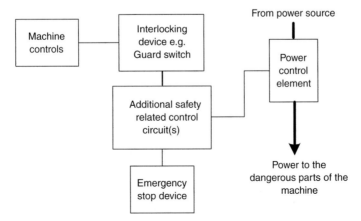

Figure 9.21a block diagram;

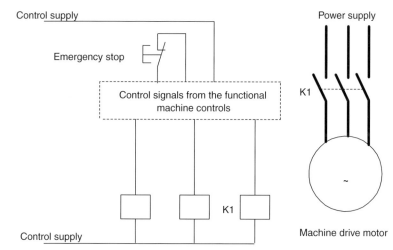

Figure 9.21b circuit diagram.

However, if the emergency arises because of a hazard which is independent of the functional safety control circuits, the emergency stop circuit can be part of the functional control circuits (*Figure 9.21*). When deciding which method to employ, all the risks from the machine powered equipment must be considered.

Wherever possible the emergency stop circuits should be hard wired and not have to rely on electronic circuits or components. However, in some proprietary safety control systems which include programmable functions, the emergency stop signals may be processed through the programmable controllers. Since these circuits are extensively monitored and include redundancy and diversity in the design to reduce the probability of failures to a minimum, they are acceptable. All electronic control systems should, as a minimum, achieve the same level of safety integrity as hard wired systems.

Chapter 10
Hydraulic safety circuits

10.1 Introduction

The use of the pressure of an incompressible fluid in transmitting force and power has many advantages over other media. While it can transmit enormous forces through static pressure loading, it has, in itself, no stored energy so that in the event of a failure all pressure is immediately lost. In the dynamic situation, it can transfer considerable levels of power from the prime source to the point of utilization while under a very fine degree of control. To benefit from these characteristics a number of conditions must be met and maintained to ensure that they can be enjoyed continuously and safely.

In safety applications hydraulic power systems have the advantage that when the supply of pressurizing oil is cut off the machine movement – whether rotary or linear – stops immediately. No brakes are required and the machine will be held motionless subject to any creepage due to leaks.

Hydraulic systems can be unique to a particular machine or be part of a factory-wide installation with hydraulic fluid piped to a range of machines from a central power source. This chapter is concerned only with those aspects of hydraulic circuits that affect the safe use of machinery. Matters concerning the provision of factory wide facilities are dealt with in *section 13.3*.

10.2 Hydraulic systems for safety circuits

For hydraulic circuits to provide the required level of safety integrity they must be assured of a consistent supply of hydraulic fluid at the requisite pressure and in adequate quantities, whether this is from an integral pump or a central source. Pressure relief valves should be fitted to ensure working pressures are not exceeded. Additional relief valves may be needed where pressure intensification may occur. Loss or reduction of pressure should cause the machine to revert to a safe condition.

The fluid should be clear of chemical contaminants, solids that may be abrasive and entrained air. Return pipes to sumps should discharge below the fluid surface to prevent aeration. Control and interlocking valves should be of a good quality and proven reliability. Pilot systems for the actuation of control valves can be either hydraulic or pneumatic – the latter having the advantage of being compact, cheap and flexible in application. Alternatively interlocking and control valves can be electric solenoid operated.

Pipework should, wherever possible, be in metal – either copper or steel depending on the application and operating pressure. Where flexible pipes are used they should be properly supported, not kinked or damaged and not rub against adjacent parts. Joints and couplings in the hydraulic circuit should be capable of withstanding the anticipated maximum pressure for the period of their economic life and not be affected by machine vibrations. Jointing compounds should be compatible with the hydraulic fluid. If the hydraulic pump is continuous running provision may be needed for cooling the fluid. Where the demand on the system is intermittent, accumulators can be included in the circuit to eliminate the frequent cutting in and out of the pump motor. Where accumulators are fitted, the supply lines to them must include non-return valves to prevent back flow to the pump and other parts of the circuit.

10.3 Hydraulic safety circuits

These circuits, using hydraulic oil as the actuating medium and working in conjunction with other safeguarding devices, provide the operators of machinery with protection from the risks created by their machines. The level of protection provided by the safety circuit should meet the requirements for the particular safety integrity level (SIL) determined by the risk assessment as being necessary for the machine. The data given in *Chapter 4* allow different methods for determining the SIL. Typical safety circuits to meet the different SILs are described in *Table 4.8*. Examples of typical basic hydraulic safety circuits are given in the following sections with each supported by a block diagram to illustrate the principles involved.

Hydraulic safety circuits should be an inherent part of the normal machine control circuits using standard 2 or 3 spool valves with appropriate porting. Actuation of the valves can be manual, solenoid, pilot oil or air operated.

10.3.1 Hydraulic single channel control interlocking

In the circuit shown in *Figure 10.1* pilot oil from the interlocking valve V1 actuates the main supply valve V2 allowing the main oil supply through to the directional control valve V3 which is controlled by the operator. Valves V1 and V2 are safety critical parts of the circuit.

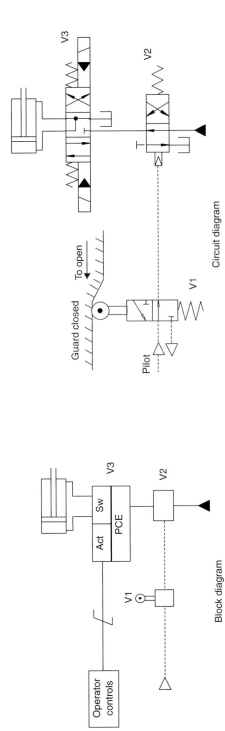

Figure 10.1 Hydraulic safety circuit with single channel interlocking and a pneumatic pilot circuit.

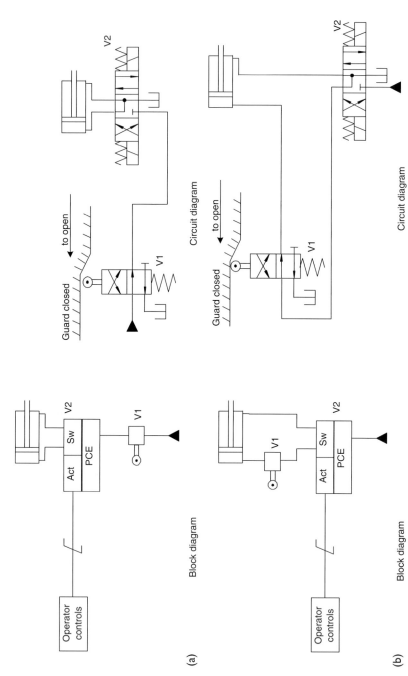

Figure 10.2 (a) Hydraulic single channel power interlocking circuit. (b) Hydraulic single channel power interlocking circuit allowing power retraction of the piston when the guard is open.

10.3.2 Hydraulic single channel power interlocking circuit

The two circuits in *Figure 10.2* show variations of a single channel power interlocking arrangement. In both these circuits valves V1 and V2 are safety critical.

Figure 10.2(a) shows a simple power interlocking circuit in which the main oil supply passes through the interlocking valve V1 to the directional control valve V2. When the guard is opened both sides of the piston are connected to dump and the piston stops moving. It may be possible to move the piston manually but on downstroking presses there is a risk of piston creepage.

In *Figure 10.2(b)* the circuit has been modified to allow retraction of the ram under power when the guard is open. This circuit should be used only on those applications where there are no hazards from the retracting ram.

10.3.3 Dual channel mixed media control interlocking circuit

The circuit, shown in *Figure 10.3* using two (mixed) interlocking media, reduces the risk of common mode failure. Two independent interlocking devices are used, the electrical interlocking switch in positive mode and the hydraulic interlocking valve in negative mode. The outputs from the two devices feed to separate hydraulic valves, V2 and V3, set in the main oil supply line to the directional control valve V4. Switch GS and valve V1 are safety related components of the circuit until one fails when the other becomes safety critical.

A variation of this circuit allows the switch GS to be connected to the operator control valve V4 such that when the guard is opened the solenoids of valve V4 are de-energized and the valve returns to its mid position. This dumps the hydraulic fluid from both sides of the ram piston and ram movement ceases.

An alternative circuit is shown in *Figure 10.4* in which the electrical interlocking is through a safety related control circuit which can incorporate the operator controls. In this mixed media circuit the positive mode electrical interlocking switch GS is connected to the electrical safety related control circuit which in turn controls the energizing of the solenoids of valve V3. The negative mode hydraulic interlocking valve V1 feeds pilot oil to the hydraulic valve V2 in the main oil supply line. These separate circuits are safety related until one fails when the other becomes safety critical.

10.3.4 Dual channel mixed media interlocking with cross monitoring

A typical interlock circuit for a downstroking single ram hydraulic press is shown in *Figure 10.5*. The positive mode interlocking switch GS energizes the solenoid of valve V2 while the negative mode interlocking valve V1 supplies pilot oil to valve V3. Both valves V2 and V3 are

158 Safety With Machinery

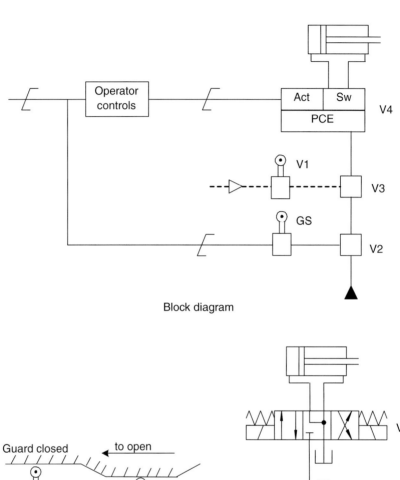

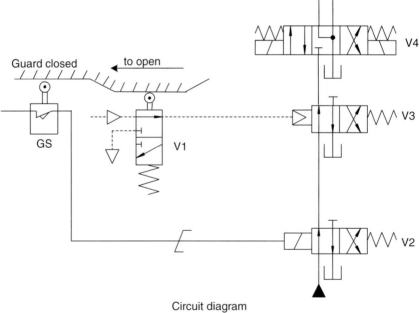

Figure 10.3 Dual channel mixed media hydraulic interlocking circuit.

Hydraulic safety circuits 159

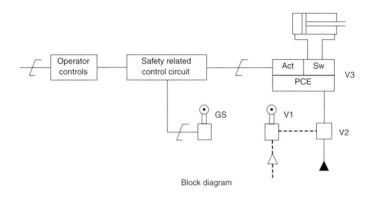

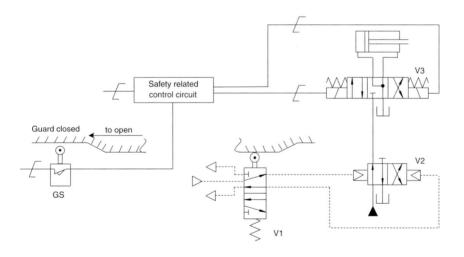

Figure 10.4 Dual channel mixed media hydraulic interlocking circuit incorporating a safety-related control circuit.

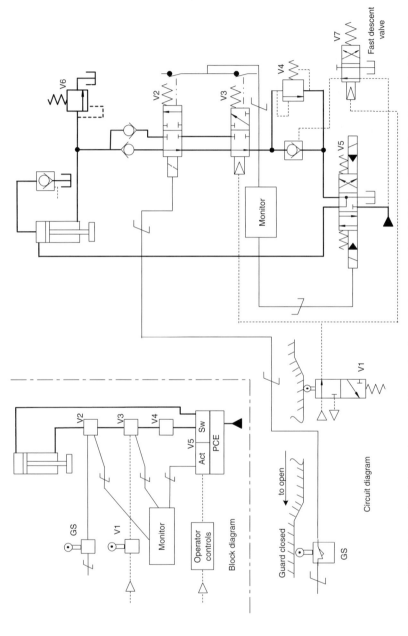

Figure 10.5 Safety circuit for a down-stroking hydraulic press incorporating dual channel mixed media interlocking, cross-monitoring and fast descent facility.

monitored and, in the guard closed condition, allow the solenoid of directional valve V5 to be energized.

The rate of descent of the ram is controlled by valve V4. The ram can be retracted with the guard open when the oil flow passes to the cyclinder via valves V2 and V3. Valve V6 is a pressure relief valve to prevent pressure intensification during the ram power stroke and is normally set at 10% above system operating pressure.

These five figures show the basic elements necessary for hydraulic safety circuits. They do not cover circuits, which can be very complex, needed for operating processes. The detailed non-safety parts of the control circuits can be developed to suit particular applications but in every case they should contain or be linked to an interlocking safety circuit of a design that is appropriate for the risks faced.

Chapter 11
Pneumatic safety circuits

11.1 Introduction

Compressed air is a convenient medium for use on both the power and control systems of machinery. It is readily available and easy to dispose of. The effectiveness of a compressed air system relies on a constant supply of air at the right pressure and in a clean and dry condition. While air is readily available and is a flexible medium for both power and control purposes, it is expensive and, because of its compressible nature, does have limitations in the accuracy of positional control.

Occasional applications arise when inert gases, such as nitrogen and carbon dioxide, are used to operate equipment but these applications are outside the scope of this book. Similarly, safety in the use of various gases as part of a manufacturing process are also beyond its scope.

11.2 Pneumatic installations

For most applications, air will be supplied to individual machines from a common main air system supplying the whole factory (13.2.1). Where a machine has its own dedicated air supply, the air should be clean and free from any form of foreign matter, dry and contain a suitable type and quantity of lubricant

Systems that use compressed air as the medium for control must either have a guaranteed source of air at a specified pressure and flow rate or be designed so that failure or reduction in the pressure or the flow rate does not increase or create new risks, i.e. does not fail to danger. Exhausts from air lines can be noisy and should be provided with silencers.

A block diagram of the basic elements of a pneumatic control system for a machine is shown in *Figure 11.1*.

11.2.1 Pressures

The pressure in the safety circuit should be such that all devices in the circuit can operate effectively. It can be either the full compressor outlet pressure or a lower pressure set by a pressure reducing valve. If the

Pneumatic safety circuits 163

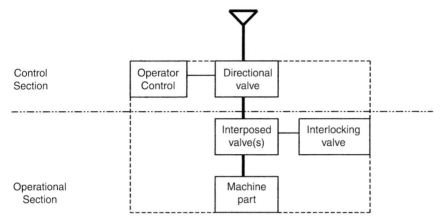

Figure 11.1 Block diagram of the basic elements of a pneumatic safety circuit for a machine.

pressure setting of the reduced pressure is critical to the safe operation of the machine, access to the pressure reducing valve controlling it should be restricted to authorized persons only.

Working pressure should be no greater than the maximum design pressure of the components to ensure they provide an economic working life and do not fail prematurely. Where there is a risk of pressure intensification a pressure relief valve should be incorporated into the particular part of the circuit to ensure that the working pressure is not exceeded.

Loss of pressure, pressure surges or intensifications should not give rise to increased risks. When the circuit pressure falls below the level necessary for the safe operation of the machine, the machine should automatically revert to a safe condition. In parts of the air system where pressure levels are critical to its safe operation, any loss of pressure should not increase the risk but cause the machine to revert to a safe condition.

If the maintenance of a supply of compressed air to certain components of a pneumatic safety circuit is critical to the safe operation of the machine, a separate air reservoir supplying that part of the circuit only should be incorporated into the overall circuit with a non-return valve in its supply line to ensure the reservoir pressure does not reduce if the overall system pressure fails (*Figure 11.6*).

Pressure locking involving equalization of the pressure on each side of the piston should not, because of the compressible nature of air, be employed on applications where non-movement of the piston is safety critical, particularly on down-stroking presses.

11.2.2 Pneumatic valves

Valve construction can be either twin spool giving two-way directional control of air flow, or triple spool with a central neutral position. For safety circuits, they should be of robust construction, of a size to suit the application, of a known quality and of proven performance and reliability.

Actuation of control and directional valves can be by:

- pilot air;
- electrical solenoid;
- solenoid control but pilot air actuation;
- return of a valve to a safe central or original position can be by one of the above three methods or by the action of return springs.

Pilot air used as the means for actuating the various operational valves can be from a separate supply or can be tapped off the main air supply system. It may be used either at full mains pressure or at a predetermined lower pressure. Arrangements should be made to ensure the selected pressure can be maintained, is not exceeded and that there is an adequate flow of air.

11.2.3 Pipes and hoses

Wherever possible the pipes of pneumatic safety circuits should be of metal securely clamped to the machine. If flexible hoses are used, they should be:

- of a construction to retain air pressure for all foreseeable configurations and flexings;
- adequately supported compatible with the range of movements;
- clear of any rubbing contact that could reduce the protective covering and strength, and
- prevented from whipping in the event of a rupture or an end fitting breaking free.

11.3 Pneumatic safety circuits

These circuits, using compressed air as the actuating medium and working in conjunction with other safeguarding devices, provide operators of machinery with protection against hazards from that machinery. In general, the higher the risk the more complex the safety circuit. The safety integrity of a pneumatic safety circuit can be improved by the use of duplicate valves (known as *redundancy*) or by the use of more than one actuating medium (known as *diversity*).

Pneumatic safety circuits should be compatible with the normal machine controls and, as far as possible, use standard two- and three-spool valves with appropriate porting. Actuation of safety-related and safety-critical control valves can be by guards, solenoids or pilot air.

A feature of pneumatic safety circuits is the use of interposed control valves, actuated by the interlocking valves or switches, positioned between the directional control valve and the machine part. If there are

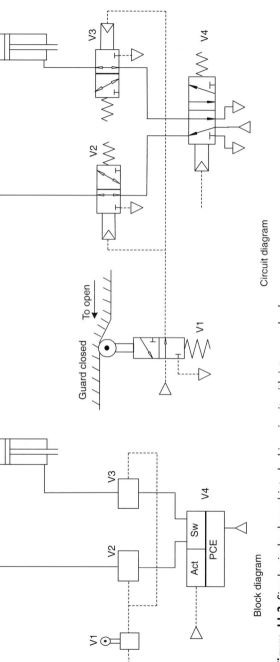

Figure 11.2 Simple single channel interlocking circuit with interposed valves.

two interposed valves they are *safety related* and if only a single valve (equalizing valve) it is *safety critical*.

The safety-related and safety-critical components of a pneumatic safety circuit should be identified during the design of the system so that steps can be taken to ensure:

- they are of known high reliability and quality;
- duplicate valves and equipment are incorporated into the circuit where necessary;
- different media are used in the interlocking parts of the circuit to reduce the possibility of common mode failures;
- where the risks warrant it the states of component valves are cross-monitored with reversion to a safe state if any fail to give the correct response.

The level of protection provided by the safety circuit should meet the requirements for the particular safety integrity level (SIL) determined by the risk assessment as being necessary for the machine. The data given in *Chapter 4* allow different methods for determining the SIL. Typical pneumatic safety circuits to meet the different SILs are described in *Table 4.8*.

Examples of typical basic pneumatic safety circuits are given in the following sections, each being accompanied by a block diagram to illustrate the principles of operation involved

11.3.1 Single channel pneumatic interlocking safety circuit

This circuit, shown in *Figure 11.2*, consists of a single positive mode interlocking valve using pilot air to actuate two interposed valves. It satisfies the requirements for an SIL of 1.

11.3.2 Single medium single channel interlocking with two opposed mode interlocking valves

The safety integrity of a single medium safety circuit can be improved by the use of opposed mode interlocking valves connected in line. The two interlocking valves are safety related until one fails when the remaining valve becomes safety critical. A typical circuit is shown in *Figure 11.3*.

11.3.3 Single channel single medium interlocking with equalizing valve

The use of an interposed equalizing valve between the directional control valve and the machine, shown in *Figure 11.4*, on actuation of one of the interlocking valves, causes the pressures on each side of the actuating piston to be equalized, pneumatically locking the piston in position. A differential pressure indicator, inserted across the feed lines between this

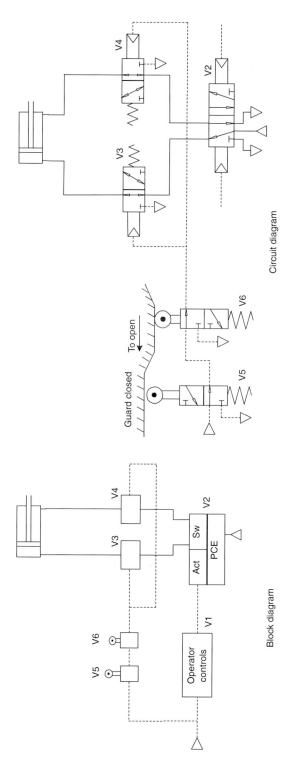

Figure 11.3 Single channel single medium safety circuit with opposed mode interlocking valves.

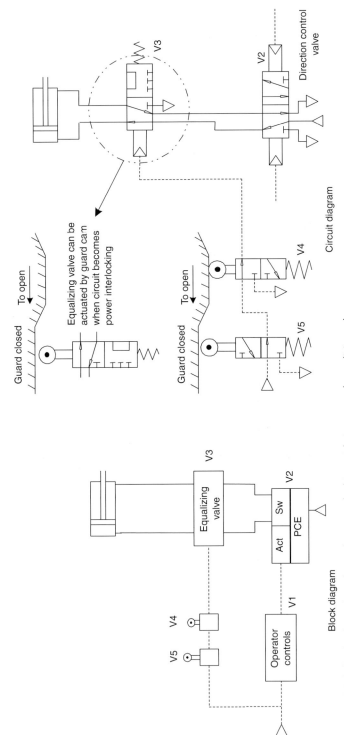

Figure 11.4 Single channel single medium interlocking with interposed equalizing valve.

valve and the cylinder, gives confirmation of pressure equalization. If the equalizing valve dumps the pressures to exhaust, the piston will stop moving but the machine part will be capable of being moved manually, However, with down-stroking machines there may be a risk of creepage.

In an alternative arrangement, the equalizing valve can be actuated directly by the machine guard when the circuit becomes power interlocking. In this latter case the equalizing valve is safety critical.

11.3.4 Dual channel single medium interlocking ciruit

The use of dual channels with opposed mode interlocking valves actuating duplicate interposed valves – known as redundancy and shown in *Figure 11. 5* – increases the safety integrity of the circuit. Because the valves use the same medium there is a risk of common-cause failure but this is offset by the greater protection provided by the duplicate interposed valves. The two interlocking valves are safety related until one fails when the other becomes safety critical.

11.3.5 Dual channel mixed media interlocking circuit

The use of dual channel mixed media interlocking eliminates the risk of common cause failure. It is usual in opposed mode interlocking for the electrical interlocking switch to be operated in positive mode and the pneumatic interlocking valve in negative mode. In this circuit the electrical interlocking signal is processed by a safety control circuit which also accommodates operator controls.

An additional feature shown in the circuit in *Figure 11.6*, that further increases its safety integrity, is the incorporation of a dedicated reservoir that causes the piston to retract under power when the guard is opened, i.e. with the interlocks actuated and power source isolated.

11.3.6 Dual channel mixed media interlocking with cross monitoring

In this circuit (*Figure 11.7*) the electrical interlocking switch GS actuates a solenoid operated valve V1 in the main supply line while the pneumatic interlocking valve V3 actuates two interposed control valves V4 and V5. The states of valves V1, V4 and V5 are checked by a monitor that is linked to the directional control valve V2 which isolates power supplies in the event of any of the valves indicating a wrong or faulty state. This type of circuit is suitable for machines requiring an SIL of 4.

11.4 Summary

These figures show diagrammatically the basic elements necessary for pneumatic safety circuits to satisfy the requirements for the various SILs. They do not show the circuits, which can be very complex, necessary for

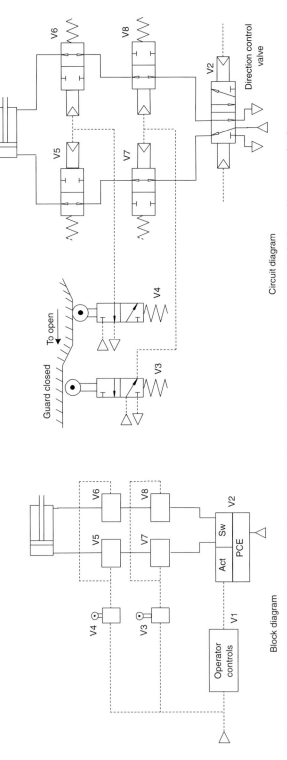

Figure 11.5 Dual channel single medium safety circuit with opposed mode interlocking valves and duplicated interposed valves.

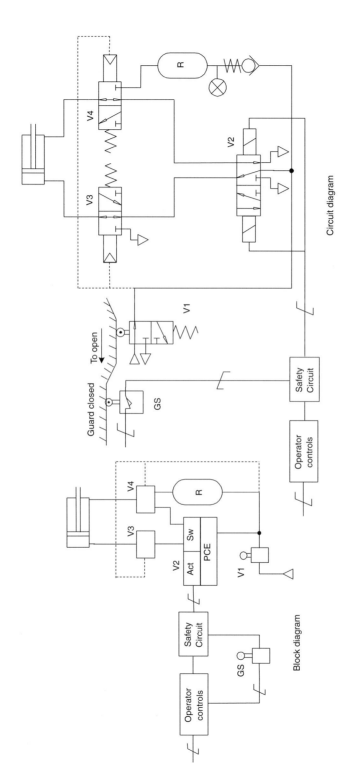

Figure 11.6 Dual channel mixed media interlocking with power retraction facility.

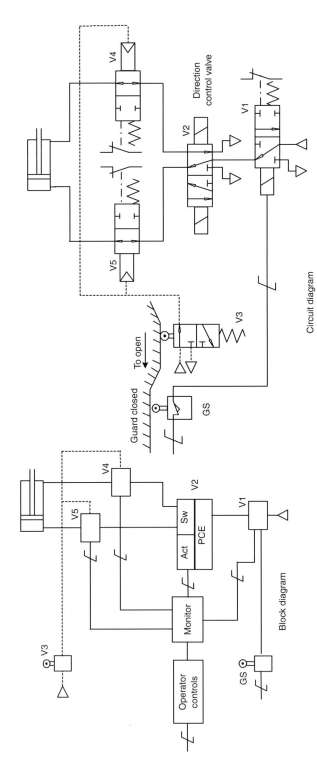

Figure 11.7 Dual channel mixed media interlocking circuit with cross-monitoring.

the operation of the machine. These safety circuits should be integrated with the normal operating control circuits to provide the desired protection for the operator.

Increasingly pneumatic safety circuits rely on programmable logic controllers (PLCs) and fieldbus systems (USBs) which incorporate extensive monitoring of the functions not only of the valves but also of the electronic components themselves and provide a very high degree of protection to the operators.

Part IV
Other safety related arrangements

The first three parts of this book have considered particular techniques that can be employed in providing protection against hazards from a range of different types of machinery. They allow the designer and engineer flexibility in selecting the type of technique to be applied to a particular machine or machinery.

However, there are some types of machines whose past record of failure and injury led to the development of specific safeguarding techniques to protect against those known, and all too frequently experienced, hazards. These particular machines have one hazard in common – they all contain stored energy whether by height or by pressure – although this is not the only hazard they present. Some of the accepted basic protective measures for these machines are considered in this part.

Other factors that influence the safe use of machinery include the manner in which they are used and the environment in which they operate. These aspects also are covered in this part.

Chapter 12
Safety in the use of lifting equipment

12.1 Introduction

Lifting occurs in almost every human activity in one form or another. It is only when the load is noticeably heavy or beyond the capacity of the individual that it becomes necessary to use lifting equipment. Over the years a vast array of different types of equipment has been developed which can range from simple items such as a builder's gin wheel hoist and the dumb waiter food lift to sophisticated remote controlled cranes capable of lifting hundreds of tonnes.

The term 'lifting equipment' covers any equipment used to lift or assist in lifting loads of any sort, including people. Broadly, lifting equipment can be split into two categories:

(i) machinery, artefacts or apparatus that provide the lifting effort or power and are referred to in this text as 'lifting machinery' which, in addition to the huge array of cranes, includes fork lift trucks, scissor lifts, gin wheels, patient hoists, passenger lifts, escalators, elevating conveyors, etc.;
(ii) 'lifting accessories' are those pieces of equipment necessary for attaching the load to the lifting machine and include slings, eye bolts, shackles, swivel bolts, clamps, magnetic and vacuum devices, lifting spreader beams, etc.

Although the processes for which lifting equipments are used are diverse and particular standards apply to specific types of lifting equipment, there are certain safety features that are common to every item of equipment used in a lifting operation.

12.2 Common safety features of lifting equipment

Every piece of lifting equipment should:

- be of adequate strength for the loads to be handled and be designed with a suitable margin for safety, i.e. an appropriate safety coefficient should be applied to design criteria;

- have guards fitted over all accessible dangerous parts;
- have spring loaded brakes on hoist drives with release by power (usually electrical) so that any loss of power causes the brakes to be applied;
- be stable when put to work, wether inherently, by the use of outriggers, by attaching to a secure base or by bracing from a stable structure of suitable strength;
- be fitted with safe load indicators incorporating overload trips;
- be clearly marked with the safe working load (SWL) or the working load limit (WLL);
- have pulleys and rope drums of diameters and design suitable for the size of hoist rope being used and be based on rope manufacturer's recommendation;
- have attached to it a plate or permanent label giving:
  - the manufacturer's name and address;
  - year of manufacture;
  - an identifying number;
  - the design load;
  - standard to which it has been designed;
- have safety clips on all hooks;
- where the lifting is by multi-rope sheaves, a hoist limit switch must be positioned to prevent sheaves making contact and overstraining the hoist rope;
- the radio of remote controlled cranes must not interfere with telecommunications frequencies, local electronic control and safety equipment or create electromagnetic pollution;
- only be operated by suitably trained individuals;
- have a schedule for inspections and maintenance;
- be inspected regularly to ensure it is safe to continue in use.

12.3 Additional features for particular lifting equipment

Lifting equipment used for specific processes or operations needs to incorporate particular design features relevant to those operations to ensure safety during use. Major specific features for particular types of lifting equipment are considered below and are in addition to the requirements listed in *section 12.2*.

12.3.1 Tower cranes

Encompass an enormous range of types across which there are a number of common features:
- the crane must be stable in use:
  - with a fixed base, or
  - fixed base and braced to a structure of suitable strength, or
  - if rail, truck or crawler mounted, be limited to the capacity of the supporting vehicle;

- every part must be permanently labelled with:
  - manufacturer's name,
  - model number or type,
  - serial number,
  - year of manufacture,
  - weight,
  - size;
- cabs must be designed to provide:
  - good visibilty,
  - adequate operator comfort,
  - suitable means of escape for operator in emergency;
- controls can be cab mounted or remote but must:
  - achieve crane movement consistent with the movement of the control,
  - be conveniently accessible and viewable by the operator,
  - contain safe load indication including, if relevant, allowance for radius of lift;
- safety limit trips should be provided for:
  - overload,
  - slewing movement,
  - luffing movement,
  - hook sheave approach to trolley sheave;
- adequate rope capacity for the maximum height of lift;
- headroom for maximum height of lift, this may be accommodated by adding sections to the tower as the height of lift increases;
- safe access:
  - to the cab by ladder with safety hoops,
  - along the jib with suitably placed life lines;
- rail, truck or crawler mounted cranes should have audible warning of approach;
- means for measuring wind speed and for securing the crane in the event of wind speeds above the safe level.

12.3.2 Overhead electric travelling cranes

- access ladders to the crane operating cabin should have safety hoops;
- operating cabins should have good visibility of the work area and be provided with heating and/or ventilation;
- means of emergency escape should be provided for the operator;
- controls must be convenient to the operator.

12.3.3 Mobile cranes

By virtue of their mobility, these cranes can create particular hazards. In addition to meeting the requirements listed in *section 12.2* they should:

- be provided with outriggers with suitable adjustable feet to ensure stability when lifting. In use outrigger feet should be placed on spreader blocks laid on well compacted ground;

- have safe load indicators and overload trips for slewing and luffing operations;
- only be used on well compacted ground;
- be positioned clear of permanent structures so that the slewing counter-weight does not create a crushing hazard.

With mobile lifting and transporting machines suitable weight, tilt or other indicators should be fitted to enable the operator to work within the limitations of the machine. Additional audible safety devices should be fitted to warn when reaching safe working limits.

12.3.4 Scissor lifts

In use the operator may have to work in close proximity to the table when there may be a risk of the lowering table trapping feet. The sides of scissor lifts should be protected by either a trip bar or a screen that fully covers each side (*section 3.3.2(g)*). When maintenance has to be carried out on the operating mechanism the lower moving rollers should be chocked to retain the table in the raised position. Note, props must not be used under the table since the table is hinged and can tilt allowing the scissor mechanism to close when hydraulic pressure is released.

12.3.5 Lifts

A lift, by definition, comprises a cage whose movement is constrained by guides and which travels between discrete platform levels or floors. Lifts should have:

- at each landing, interlocked gates that cannot be opened except when the cage is at the landing and the cage cannot be moved from a landing until the gates are closed;
- a notice prominently displayed in the cage indicating the maximum safe load;
- dual support cables each capable of carrying the full load;
- arrester gear, independent of the support cables, to bring the cage to a halt in the event of a support rope failure;
- trip device to prevent cage movement if the arrester gear is inoperative.

Where the lift carries passengers additional safety features should include:

- a gate to the cage interlocked to prevent lift movement when it is open and which cannot be opened until the cage is at the landing;
- control gear that ensures the cage floor stops level with the landing floor;

- landing and cage doors should have trip devices fitted to the leading edges to prevent trapping in closing doors;
- the provision of adequate lighting in the cage;
- means for emergency communication that will work in the event of a power failure;
- lifts operated by hydraulic power must have a safety locking valve fitted onto the lifting cylinder such that it locks the hydraulic cylinder on failure of hydraulic pressure;
- Paternoster lifts must be fitted with trip bars or edges at the top of the access opening of each landing and each cage. The trip arrangement must ensure the lift is brought to a stop before injury can be caused.

12.3.6 Fork lift trucks

Fork lift trucks should be provided with:

- overhead and rear guards to protect the operator;
- means to immobilize the truck when the operator leaves it;
- a pressure locking valve fitted onto the hoisting cylinder that locks the oil in the cylinder in the event of oil failure;
- a plate indicating the safe loads at different positions on the forks.

12.3.7 Car hoists

Hoists for raising cars to enable work to be carried out underneath them should have:

- a lifting arrangement that ensures that the vehicle is raised on a level keel;
- scotches or deadlocks that retain the hoist in its raised position;
- means to control the rate of descent of a loaded hoist.

12.3.8 Escalators

Escalators must be provided with:

- means to ensure that children and others with small feet cannot become trapped at the leaving point;
- side skirts along the whole length designed to prevent trapping between the moving steps and side frame members;
- emergency stop buttons prominently located at top and bottom (entry and exit) of the escalator;
- suitable access to below the moving steps to permit regular servicing and cleaning;
- fixed automatic fire fighting system below the moving steps.

12.4 Lifting accessories

Lifting accessories covers a vast range of items for joining the load to the lifting equipment. Some have general application to a wide range of lifting situations while others are of a very specific nature to meet one application only. However, they all have a number of common features in that they must be:

- permanently marked with safe working load;
- of adequate strength;
- inspected regularly for signs of damage or deterioration. (A brief visual check every time they are used can identify faulty items.);
- maintained in a good condition;
- stored in a proper fashion and under conditions that prevent damage and deterioration;
- withdrawn from service if they show signs of damage or deterioration.

Lifting accessories, which include slings (of chain, wire, fibre rope, webbing and polyester strands), eye bolts, bow nuts, shackles, lifting beams, etc., should only be used within the load limits set by the manufacturer. Particular care needs to be taken when using eye bolts and bow nuts to ensure that the thread of the bolt matches the thread of the tapped hole in which it is inserted.

12.5 Circumstances requiring special precautions

In the use of lifting equipment circumstances arise that require special precautions to be taken. These may or may not be related to the operation of the lifting equipment but can include:

- a hostile operating environment, such as corrosive fumes, high temperatures, dust laden atmospheres, etc., likely to cause accelerated deterioration of equipment requiring an increased safety coefficient to be applied to the design and greater frequency of maintenance and servicing;
- the use of remote or radio controls which must not create electro-magnetic contamination nor emit radiations that interfere with the functioning of adjacent equipment;
- special purpose lifting accessories for attaching the load to the crane which must be stable in use and be designed with an adequate safety coefficient;
- access for operators and for maintenance. If by vertical ladders they must be fitted with safety hoops;
- controls must take account of ergonomic principles in their design and layout;
- cabs of cranes operating in noisy atmosphere should be sound proofed and have effective means of communication between the operator and banksman.

12.6 Precautions when handling lifting equipment

12.6.1 Stability

All major machine components should be designed so as to be stable during manufacture and assembly and the finished machine should be capable of withstanding reasonably anticipated internal forces from its operation and, where pertinent, external forces such as wind pressure, vibration, impact, tension when connected to adjacent machines, etc. Where the functional configuration results in a reduced stability, the required stability can be achieved by broadening the base of the machine, bolting the machine securely to a solid foundation, the addition of appropriately located supporting members or struts, or other suitable means.

12.6.2 Lifting

Components of the machine or the machine itself which are too heavy to be handled manually should be provided with suitable means for ensuring they can be lifted safely. These can be built-in lifting lugs, tapped holes for lifting eyes or other means for the safe connection of the lifting sling or hook to the machine or part. The weight of the part to be lifted should be marked clearly on the part and where the load is eccentric, the position of its centre of gravity should also be marked. Any special lifting requirements should be detailed in the operating manuals. Lifting equipment that is attached to a machine should be inspected regularly.

12.6.3 Handling

Machines, components, fixtures, attachments, etc. that need to be handled manually should be of a size, weight and configuration to allow this to be done safely. Where this is not possible the design must allow for mechanical handling.

12.6.4 Transporting

The machine, its component parts, fixtures, etc., should be capable of being transported safely. Care must be taken to allow for the dynamic effects of movement particularly turning, accelerating and stopping.

Chapter 13
Safety with pressure systems

13.1 Introduction

Since the earliest days of the Industrial Revolution the use of fluids as a means for transmitting power has been recognized and exploited. Initially through the use of water power then, as technology developed, through the use of fluids – mainly water, steam and air. This chapter considers the safety implications of the use of the three main pressure systems found in industry today – air, hydraulic and steam.

Each system has its advantages and disadvantages which need to be set against the benefits it brings. The use of compressible fluids as the means for transferring power, whether as heat or contained energy, has great application in industry. However the price of convenience is high and there are many hazards and risks associated with the energy stored in compressed fluids. On the other hand, incompressible fluids – oils and water – have the ability to transmit great power with the safety benefit that in the event of a failure of the pressure containment there is no stored or potential energy.

There are factors common in the design of all plant handling fluids at pressure that are aimed at ensuring its safe use:

- containment vessels must be designed and constructed to accepted standards and codes and have an adequate margin of safety in operation;
- provision must be made to:
  - prevent over-pressure;
  - display clearly the maximum safe working pressure;
  - show the contained pressure;
  - identify individual vessels and equipment;
  - prepare a scheme for periodic internal and external examinations;
  - display information on design and construction standards;
- those sections of a pressure sytem operating at reduced pressure should be protected by a suitably set pressure relief valve.

The safety aspects of systems for the supply and use of the most common media, air, hydraulic fluids and steam, are considered below.

13.2 Compressed air

Air is a very convenient medium for transmitting power since it is freely available and easily disposed of to the atmosphere without causing pollution, but it is expensive to produce and can be dangerous if misused.

Because of its compressible nature, it can create hazards through the energy stored in it, but by following well established and proven procedures it can be utilized safely and effectively. Hazards can occur in an installation from:

- leakage through a pinhole creating an air jet that can be intense enough to puncture the skin;
- the build up of oil and carbon deposits inside the pressure containment can result in fire and explosion;
- dirt and oil in the air can block valves and prevent them from working properly;
- over-pressurization due to faulty relief valves giving rise to a risk of pressure vessel rupture;
- noise from unsilenced exhausts.

Air can also be evacuated to form a vacuum, which has industrial application but presents particular hazards.

13.2.1 Air system services

In the design of a system for the provision of air supplies, account should be taken of:

- air volume demand at each location to be served;
- air pressure required at each location;
- provisions for isolating individual machines ;
- provisions for isolating sections of the system;
- provision of suitable filters, drains and water extractors;
- means for supporting the pipe runs;
- routing of pipes to avoid access ways;
- identification of the air pipe lines (*Appendix 5*);
- the need to maintain the quality of the air up to the point of usage.

Once the overall parameters of the system have been established a schematic diagram of the system can be developed similar to that for a typical small to medium sized air service system shown in *Figure 13.1*.

Schedules should be prepared for lubrication, servicing, maintenance and for periodic inspection and examination of the system.

If the air is to be used for breathing, such as in certain personal protective equipment, rescue breathing apparatus and pressurized working, an additional special filter should be provided to remove oil vapour.

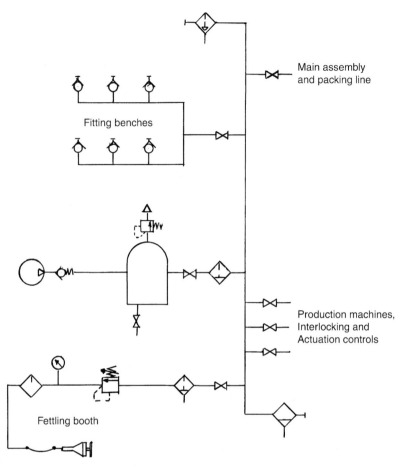

Figure 13.1 Schematic diagram of a small to medium air line system incorporating safety features.

13.2.2 Compressor

The size and output of the compressor must match the current and foreseeable air demands of the system. The process of compressing air generates noise so the compressor should be either a packaged unit contained within its own sound proof enclosure or be located away from the workplace in its own sound proofed room. Also the fact of putting work into the air to compress it raises its temperature. In both cases adequate ventilation must be provided to supply both the service requirements of the air system and meet the cooling needs of the motor and compressor. Facilities and space should be allowed in the enclosure for carrying out maintenance.

Compressors, such as positive displacement compressors, that may generate vibrations should be installed on vibration proof mounts. For small installations, free standing screw type compressors, which work to

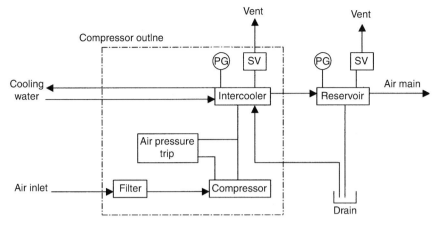

Figure 13.2 Block diagram of a single stage compressor.

demand and do not require an air receiver, can be located in the workplace since they are quiet.

A block diagram of a typical single stage compressor is shown in *Figure 13. 2*.

Compressors should feed directly into an air receiver of suitable size to accommodate fluctuations in the system demand. A heat exchanger may be required to maintain the compressed air temperature below the maximum allowed in the receiver. Failure to do this could result in the temperature rising high enough to ignite any oil/air mixture resulting in an explosion or internal fire.

Compressors should comply with EN 1012 – 1 and be provided with:

- adequate guards over belt and shaft drives (F*igure 3.3.3.d*);
- suitable safety (pressure relief) valves at intercoolers, coolers and reservoirs;
- pressure gauges at compressor outlet, on each cooler, intercooler, reservoir and on pressurized lubrication systems;
- temperature gauge at compressor outlet;
- on water cooled compressors, a water temperature trip at water outlet;
- lubricating oil level indicator;
- air inlet filter with indication of pressure drop across it;
- a well ventilated site having a clean cool air supply free from flammable or corrosive contamination.

13.2.3 Vacuum pumps

The exhaust from vacuum pumps is noisy and small units located in the work area should be fitted with suitable silencers. For larger installations and those serving a number of machines the pumps should be located in their own separate sound proofed room with restricted entry requiring the wearing of hearing protection. Vacuum pumps should comply with EN 1012–2.

Safety With Machinery

The belts and shafts driving vacuum pumps should be guarded and a vacuum gauge fitted at the inlet to the pumps. They should be provided with an oil level indicator and a label specifying the grades of oil to be used.

13.2.4 Air receivers

Air receivers should:

- be clearly marked with their safe operating pressures ;
- be of a size to match the compressor output and accommodate fluctuations in air demand;
- carry a label giving:
  - the manufacturer's name,
  - a serial number,
  - date of manufacture,
  - the standard to which the vessel was designed and built,
  - the maximum design pressure (or vacuum), and
  - the design temperature.
- be provided with:
  - an inlet and outlet port,
  - a pressure gauge,
  - a pressure relief valve,
  - a drain,
- a manhole to permit internal inspections;
- have a written schedule for inspections and examinations if the product of pressure and volume is greater than 250 bar litres.

13.2.5 Air driers

Air driers may be necessary where:

- the moisture content of intake air is high;
- pipelines pass through an area of low temperature;
- very dry air is required by the production process.

For normal industrial use, centrifugal water separators should be installed at suitable positions in the pipeline. Refrigerant driers can be used to reduce the dew point of the compressed air to below that of the lowest likely temperature along the pipeline. For very dry air, absorbent devices will be required.

13.2.6 Pipework

Installed pipeline systems should:

- be of adequate strength to contain the maximum foreseeable working pressure;
- be of adequate size to cope with the current and likely future demands without causing excessive pressure drop – typically not more than 5% of pressure required at outlet point or 0.3 bar whichever is the lesser;

- be of steel pipe, adequately supported and not interfere with access or gangways;
- slope away from the supply point, i.e. fall in the direction of flow, with a drain at the lowest point;
- incorporate valves in the main pipeline and in each branch pipe to allow sections to be isolated;
- be routed so as not to pass through cold areas likely to cause condensation. If this is not possible, the pipeline should be adequately lagged for the relevant length;
- incorporate suitably placed drains and water traps;
- be clearly identified by colour coding (*Appendix 5*).

The air system to each machine should contain its own isolating valve, moisture extractor and drain which ideally should be automatic, lubricator and pressure indicator. Care should be taken to protect the glass bowls attached to drains, filters and lubricators and to ensure they are correctly fitted before being pressurized.

Pipe connections to the main airline should wherever possible be permanent. Where quick release connections are used they should be of adequate size for the flow required, lock together without the risk of becoming disconnected and be simple and quick to release without requiring excessive force. They should be positioned to point downwards so they do not collect dust and dirt.

Where flexible pipes are used and failure could give rise to a risk of the loose end whipping, the pipes should be suitable clamped. The material of flexible pipes should be resistant to any oils, solvents or other liquids they may be in contact with in the course of operation of the machine. Care must be taken to ensure that moving parts of the machine do not rub against any part of the pipework and that flexible pipes cannot become trapped between moving parts of the machine. Second-hand flexible pipes should not be used nor should flexible pipes with kinks. Where the pipes are flexed in operation, the minimum radius of flexing should not be less than maker's recommendation. Pre-assembled flexible hoses should be marked with their date of manufacture.

13.2.7 Components

All components in a pneumatic system should be rated for the working pressure of the system to ensure they give an adequate working life. Each should carry a label stating:

- maker's name and address;
- maker's product identification;
- rated operating pressure;
- symbolic diagram of the item's function;
- for hose assemblies, the date of manufacture.

Before assembly, all components should be clean, clear of swarf, scale, burrs, etc. and all passages, holes and ports clear of any materials likely adversely to affect their functioning.

190   Safety With Machinery

All components should be mounted securely and in a manner that will prevent their becoming loose in operation. The mountings of all components converting air pressure to mechanical power or force, such as cylinders and motors, should be capable of withstanding all foreseeable operating forces over the expected working life.

Exhaust outlets from motors, cylinders, etc., should be fitted with suitable silencers so they do not present a noise hazard to operators or others in the vicinity.

13.2.8 Maintenance

Pneumatic circuits should be inspected regularly for signs of leakage, wear and deterioration in performance. Components should be serviced in accordance with the maker's recommendations. Filters, drains and lubricators should be checked daily.

If components are removed for repair or maintenance, care should be taken to ensure that no foreign matter enters or remains in the device. Replacement components should be the same as the original or have the same operating characteristics.

13.3 Hydraulic installations

The components of a hydraulic safety circuit require a constant supply of oil at a particular pressure and in a condition that will not affect adversely the operation of those components. This can be provided either by an installation unique to a particular machine or by a central installation supplying oil to a number of machines. Hydraulic oil can be used as the medium for power transmission and as pilot oil for control circuits.

13.3.1 Pressures

All parts of the system should be designed to withstand the maximum operating pressure, whether of the pump, the supply main or the pressure at which pressure control devices are set.

Working or rated pressure of the system should be set sufficiently below the maximum design pressure of components in the system to ensure that they give the required performance over an acceptable working life. A pressure relief valve should be incorporated into the circuit to ensure the set working pressure is not exceeded.

Where pressure surge or intensification may occur and could become a hazard by creating pressures in excess of the working pressure, suitable pressure relief devices should be incorporated at the appropriate positions in the circuit (*Figure 10.5*).

Loss of pressure, pressure surges or intensification should not give rise to increased risks. Where pressure levels are critical, any fall of pressure below the critical level should cause the machine to revert to a safe condition. In safety-critical lifting operations – fork lift trucks, passenger lifts, scissor lifts, etc. – protection against the effects of failure of pressure in the supply pipe line should be provided by hydraulic locking valves attached to the operating cylinder(s) which lock the pressure in the cylinder when the control pressure falls at a greater than predetermined rate.

Single ram down-stroking presses should have a hydraulic locking valve fitted to the lower inlet to the cylinder body to prevent inadvertent down-stroking in the event of oil supply loss or failure.

The control circuits should be arranged so that failure of the pilot pressure causes the machine to go to a safe condition.

13.3.2 Hydraulic fluid

Mineral based hydraulic oils are relatively cheap but are flammable and a fine leak at pressure can create an explosive mist. Synthetic based non-flammable hydraulic oils are more expensive but do not give rise to the same fire risk. However, they do contain additives that can be hazardous to health so suitable precautions should be taken when handling them and when clearing spillages and leaks.

The various materials used in circuit components should be compatible with the hydraulic fluid and should not deteriorate within their expected life span.

To ensure that the temperature of the hydraulic fluid does not exceed the maximum stated by the supplier, suitable coolers may need to be included in the hydraulic circuit.

Included in the pipe system should be suitable filters, water separators, coolers, etc. to maintain the purity of the hydraulic fluid. Where contaminants do enter the fluid, their effects should be checked and if found to be deleterious, arrangements made to remove the contaminant or to replace the fluid.

Any leakage from a component or joint should be collected in a suitable receptacle or drip tray until a repair can be effected. Random leaks can cause a slipping hazard and the affected floor area should be covered with a suitable absorbent material and cleaned up as soon as possible.

Hydraulic fluid should be checked regularly to ensure its condition does not deteriorate due to water, air or other contaminant and to maintain the level in the sump.

13.3.3 Pumps

The pump performance should be such that, at the working output pressure and flow, it will remain serviceable over its expected working life.

In systems where the pumps run continuously they should be provided with pressure relief by-pass of a suitable capacity to accommodate the full pump output to absorb fluctuations of demand by the system. In other systems where the pumps run only on demand, i.e. fork lift trucks, scissor lifts, security gate with power opening, etc., pressure relief valves should be fitted to ensure the working pressures do not exceed the design maximum for the system.

With hydraulic systems it is essential that any entrained air is removed before the fluid is drawn into the pump. Failure to do so will result in cavitation in the pump with the generation of high noise levels, possible mechanical damage to the pump and failure of the pump to meet its required performance. Restriction of flow at the pump suction inlet should be avoided since it can result in cavitation

The outlets of all return drains to the sump should discharge below the oil surface to reduce the possibility of aeration.

13.3.4 Pipework

Pipework must be designed to withstand the working pressure of the system and be of a size to accommodate the volume flow required by the devices served without creating an excessive back pressure. It should be colour code marked (*Appendix 5*) to identify its contents. All pipes in a system should be clean and free from foreign materials, swarf, burrs, scale, etc.

Hydraulic pipe installations should be of rigid steel with adequate supports. Where flexible pipes are used they should be capable of withstanding the maximum operating pressure, not be kinked or stretched and be restrained by rigid supports at each end. Couplings and joins in the pipes should be suitable for the system pressure and should be capable of withstanding any vibrations – both mechanical and hydraulic – likely to be met in service. Pipework should be labelled to indicate its function in the circuit.

Circuit design should allow for individual components to be removed for maintenance, with suitable arrangements to isolate appropriate parts of the circuit. Suitable separate drain and bleed points should be provided to allow the system to be drained and any trapped air removed but should be arranged to ensure that air cannot enter the system.

Gas loaded hydraulic accumulators should use only an inert gas, such as nitrogen, and be arranged with a non-return valve in the feed line so that when the pump motor is switched off the accumulator maintains its pressure. There should be an impermeable membrane between the hydraulic oil and gas spaces.

13.3.5 Instruments and controls

Instruments to measure, indicate and record conditions in the hydraulic circuit should be provided at various points in the system sufficient to enable the operator to control the system safely. Such instrumentation should be located in easy view of the operator. On complex systems automatic controls may be necessary to ensure the system operates within its required parameters.

Where pressure levels are critical to the safe operation of the system, the devices controlling that pressure should be accessible only to authorized operators. Similarly where adjustable pressure relief valves are installed to prevent pressure intensification or other pressure build up, access to the adjustment of those valves must be restricted.

The controls that have to be manipulated should be located where the operator is not put at risk from the hazards of the machine.

13.3.6 Components

Each component in a hydraulic system should be rated for the system operating pressure, be of a design to ensure effective service over an anticipated life and should carry an identifying label stating its operating limits. Components should only be used for the function for which they were designed.

Where components perform a safety-related or safety-critical function they should be of a suitable quality, proved reliability and be monitored.

Before being installed in a circuit, all components should be clean and free from swarf, moulding sand, scale, burrs, etc., and all passages and holes clear of any material likely to block them.

All components should be mounted so that they are secure. Where it is necessary for the operator to view them, they should be within easy visual range of the operator. Pumps and components for transmitting power should be securely mounted so that in normal service they remain rigidly fixed. Such mountings should be of adequate strength to resist any operating forces likely to be met in service.

13.3.7 Maintenance

Hydraulic systems and components should be maintained in accordance with the manufacturer's instructions. Care must be taken during maintenance that components do not become contaminated and that they remain free from foreign matter, swarf, etc. Replacement components should be the same as the original or have the same performance characteristics.

13.4 Steam

Steam as a source of power has been used in industry ever since the start of the Industrial Revolution in the 18th century. Then little was known of the hazards associated with it and how to avoid them with the result that boiler explosions were a common occurrence and extracted a high toll in deaths and injuries. Many of the hazards remain but modern technology and practices have reduced their effects to a minimum. This section deals with the safety devices and practices currently incorporated into modern steam generators to ensure their safe operation.

A block diagram of a typical steam generator is shown in *Figure 13.3*.

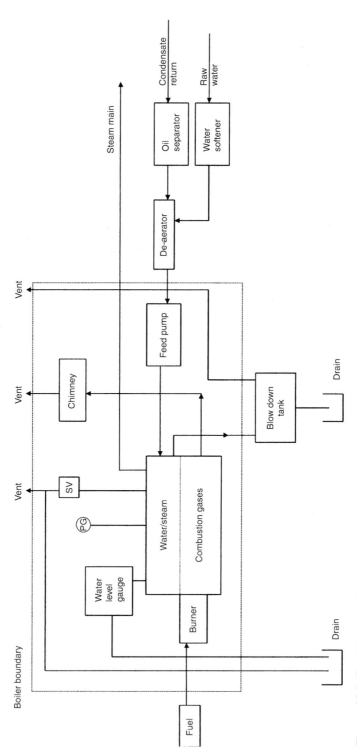

Figure 13.3 Block diagram of a typical steam generating plant.

13.4.1 Siting

The steam generator should be housed in a separate building or separated from other parts of a building by a fireproof wall and be provided with at least two means of emergency escape. The location may be dictated by circumstances, local planning controls or environmental considerations or a combination of these. It should be arranged so that long steam pipe runs are avoided, fuel storage facilities are conveniently at hand and there is an adequate supply of suitable water.

13.4.2 Steam generator construction

Modern steam generators tend to be of welded construction and their designs should conform with the legislative requirements and standards relevant to the type and operating conditions. However, many older steam generators still remain in active service. There are a number of safety features that should be common to all steam generators which should have:

(a) a pressure gauge to show the steam pressure inside the boiler with clear marking to show the maximum safe working pressure;
(b) a safety valve set to lift at working pressure plus 10%;
(c) gauges to indicate the level of water in the steam generator;
(d) facilities for blowing down the residues that accumulate in the water within the steam generator;
(e) an alarm to warn of low water levels;
(f) a scheme for their regular inspection to check for signs of wear and damage to steam generator components and that the safety devices operate satisfactorily.

The steam generator should be sized to give the highest efficiency of operation at the foreseeable steam demand. Each steam generator should carry a plate giving its design and working pressures, maker's name and address, date of manufacture and the code to which it was designed.

13.4.3 Fuel

Three main categories of fuel are used; solid, gaseous and liquid. Electricity is also used in electrode steam generators.

13.4.3(a) Solid fuel

In the form of coal, coke, wood, peat, refuse and other waste products. Problems of storage and in handling when dusts can be generated that can create clouds of explosive mixtures. Handling in modern steam generators is largely automatic with fuel being delivered by conveyors which should be guarded (*section 3.3.3.c*). Ash produced by the combustion process has to be removed and disposed of.

13.4.3(b) Liquid fuel

Easier to handle than solid fuel, it is stored in either horizontal tanks, above or below ground, or in very large over-ground vertical tanks. In both cases the tanks must:

- be of adequate strength;
- include manholes and inspection openings;
- be surrounded by a bund that can contain 110% of the capacity of the tank or tanks if more than one. The bund walls must be constructed to contain any leakage or spillage as regards both strength and chemical resistance and must not have any holes or outlets. Removal of rainwater should be by pump from a sump located within the bund;
- be provided with filling connection points which should be within the bund wall;
- for heavy oils, be provided with suitable means for heating;
- not be completely filled but left with an ullage space of about 10% of capacity;
- be provided with indicators of the level of the contents;
- have a rain proof vent;
- be provided with a fire stop valve on the outlet pipe.

13.4.3.3 Gaseous fuel

The gaseous fuels used are mainly natural gases and can include methane from waste sites. Some small boilers may be fired by liquified petroleum gases (LPG). LPG in liquid form can be stored in a suitable local tank or in its gaseous form piped direct from the supplier. Safe handling relies on the integrity of the pipework system. The design and installation of the pipework may be subject to local legislative requirements.

The main hazard is from leaks which, since many of the gases are heavier than air, tend to collect in low places and can form explosive mixtures with air. The increasing availability of gas supplies and the convenience of it as a fuel encourage conversion from coal and oil firing. Care should be exercised to ensure in each case that the boiler is suitable for conversion particularly in the case of multitubular shell boilers.

13.4.3.4 Combusters

Combusters are the means by which the fuel can be burnt in the steam generator. They can be:

- jetted burner for oil, gas and pulverized fuel;
- static or moving grate for solid fuels;
- fluid bed for solid fuels.

Combusters should have facilities for adjustment to allow the correct setting for efficient firing and hence minimal environmental pollution.

Assisted air supply may be required either by induced draught with air drawn through the combuster by a fan at the chimney end or by a combination of forced and induced draught where air is blown in at the combuster and the products of combustion are drawn from the steam generator by a second fan and blown up the chimney stack. In both cases the drives to the fans should be fully guarded (*section 3.3.1f* and *g*). Forced and induced draught fans should be run for some minutes before fuel is introduced to remove any residual combustible gases. Induced draught fans need to be designed to cope with hot exhaust gases and possible abrasive ash.

Combusters should be designed to allow easy maintenance of:

- jet burners – the cleaning of the jets;
- moving grate burners – subject to high degree of wear and provision should be made to allow replacement of components;
- fluidized bed burners – subject to abrasion from the sand of the bed and provision should be made for replacement of components.

13.4.4 Steam generator and steam supplies

13.4.4(a) Feedwater

The quality of water that a steam generator needs depends on its type and output. The higher the output (pressure and steam quantity) the higher the quality of feedwater needed.

Use of untreated water can result in:

- scale or deposits on heat exchange surfaces causing overheating and early failure;
- corrosion and consequent weakening of pressure parts;
- oxygen pitting with the development of blow holes;
- carry-over or priming where the chemicals in the water froth up and are carried over into the steam pipes.

The treatment and conditioning of feed water is essential in preventing premature failure of the steam generator and associated pipework since it:

- reduces scale deposits on steam generator heating surfaces, preventing overheating and improving its efficiency;
- prevents corrosion of the steam generator, steam and condensate systems;
- prevents carry-over and priming.

Feedwater derives from two sources:

- condensate from used steam but this may have been contaminated by the process in which it has been used;
- raw or make up water from an external source which may contain unacceptable levels of dissolved chemicals.

Provision should be made for checking condensate for contamination and for treating it. Depending on the type of contamination, the condensate may need to be chemically treated for soluble contaminants or passed through a separator for insoluble contaminants, or both.

Make-up water, from either water mains or natural source – pond or river – may contain dissolved chemicals that could interfere with the safe operation of the boiler. Provision must be made for suitable treatment processes such as:

- base exchange (strong acid cation exchanger);
- dealkalizer (weak acid cation plus degasser);
- two stage demineralization (de-ionization);
- evaporation;
- reverse osmosis;
- electrodialysis.

The latter four processes result in better quality water. Ideally water should be slightly alkaline with a pH value of 11. A de-aerator may be needed in the feedwater treatment to remove oxygen.

13.4.4(b) Blowdown

Steam boilers should be blown down to remove suspended and dissolved solids that have built-up in the water as a result of evaporation. Blowing down can be performed intermittently, continuously or a combination of both.

Blow down should be to a special blow down tank which is a pressure vessel designed to withstand 25% of maximum boiler pressure. The tank capacity should be sufficient to allow flash-off of the steam which should be vented in a safe manner to atmosphere through pipes of a size not to create a back pressure. Blow down condensate can be used to save heat by preheating feedwater and when cool enough can be discharged to drain. Blow down pipework should be sufficiently robust and well supported to withstand rapid pressurization, thermal shock, high velocity flow and vibration.

If the blow down tank serves more than one steam generator, only one blowdown key should be available which when in use must not be capable of being removed from the blowdown valve. Where the blowdown valves of a range of boilers require different sized keys, all the keys should be joined (welded) together so they cannot be separated. Each blowdown pipe, in addition to the blowdown valve, should incorporate either a screw down non-return valve or a non-return valve plus a separate isolating valve.

13.4.4(c) Steam supply and distribution pipework

Steam is delivered to where it is needed through a pipework system which must be designed to ensure that when delivered the steam is the right pressure, quantity and quality for the process. The design of pipework needs to take account of:

- steam demand at the various factory locations served;
- steam pressure required at each location;
- the dryness of steam required;
- provisions for isolating individual plant and machinery for inspection, maintenance and repairs;
- facilities to isolate steam mains for inspection, maintenance and repair;
- collection of condensate from each isolatable section and from appropriate points in the pipe run and its return to hot well;
- means for supporting the pipes;
- the provision of lagging and its protection;
- the thermal expansion of pipe runs;
- routing of pipe runs to avoid interfering with access ways;
- the identification of steam and water pipelines (*Appendix 5*).

Piping should be sized for a steam velocity of 25 – 30 m/s. Piping that has to be pressure tested should be provided with vents at its highest points. Where more than one boiler feeds into a pipework system, in additon to the stop valve mounted on the steam generator, the steam feed pipe from each steam generator must be fitted with either a globe stop and check valve with facility for being locked in the closed position or a stop valve capable of being locked in the closed position and a separate check valve.

Pressure reducing valves should be used where a process or machine requires a steam supply at a pressure lower than the mains pressure. The pressure system downstrem of the reducing valve must be protected by a safety device set for the reduced pressure.

Except for power generation, steam is normally supplied as saturated steam with a small water content. If dry steam is required a suitable water separator should be included in the pipe line.

Pipework should be fitted with manual drains and steam traps to prevent the build up of water pockets which can cause water hammer. This is particularly important at start up before the pipes become hot. Steam valves of grey (flake graphite) cast iron are vulnerable to water hammer.

To allow for the movement of steam pipes due to thermal effects the supports can be:

- anchors which fix the pipe solidly, allow no movement and become points from which linear expansion takes place;
- guides which allow axial movement but no lateral movement;
- hangers which allow both axial and lateral movement while supporting the pipe. Hangers can have solid support rod suspensions or incorporate a spring device which allows limited vertical movement.

Expansion movement in the pipe can be accommodated by strategically placed right angle bends, by special expansion loops or by expansion bellows but the latter require special mounting to prevent distortion and failure.

13.4.5 Emissions

Flue gases from the combustion chamber are exhausted to atmosphere through the chimney stack. Depending on the fuel burnt, the gases will contain a range of environmental pollutants. Emissions of these pollutants to the atmosphere are restricted by international legislation requiring treatment of the gases before they can be released. The main pollutants are:

- carbon dioxide which is produced when carboniferous substances are burnt and is one of the main 'greenhouse' gases responsible for global warming. Carbon dioxide emissions can be reduced by:
    - utilizing a fuel with lower carbon content such as natural gas;
    - improving the combustion process through the use of more efficient burners;
- sulphur dioxide from the burning of coal and oil which both contain sulphur. This reacts with moisture in the atmosphere to form sulphurous acid (H_2SO_3) the so-called acid rain;
- various oxide of nitrogen (referred to as NO_x) from the nitrogen in the air – it comprises 75% nitrogen. This also contributes to the acid rain;
- water is formed by the combination of hydrogen and oxygen and is not a pollutant but it can form a plume under certain conditions to give the impression of a smoking chimney;
- various dusts from coal and solid waste matter;
- traces of various heavy metals from the burning of domestic and industrial waste.

Removal of these pollutants is necessary and can be achieved by:

- washing or scrubbing the combustion gases through water sprays;
- chemical treatment to capture the gases;
- mechanical filtration of larger dust particles;
- use of electrostatic precipitators to collect the finer dusts.

Ultimate emission into the amosphere may be subject to strict controls with the height of the chimney being dictated by environmental conditions and the velocity of gas at discharge being specified so as to prevent plume inversion to the downwind neighbourhood.

Chapter 14
Safe working with equipment

14.1 Introduction

Ideally, all work with machinery, plant or equipment should be carried out in a safe manner and follow the instructions given during training and in the manufacturer's operating instructions. However, any operation of machinery and equipment is potentially hazardous but the hazards need not give rise to injury or damage if the safe working rules are followed and common sense applied. Not all operating circumstances can be foreseen at the design stage and additional precautions found necessary by operating experiences may be needed. There are a number of procedures and practices which, if followed, contribute to ensuring safety when working with machinery. These are discussed in the following sections.

14.2 Systems of work

A safe system of work is a considered method of working that takes full and proper account of the hazards created by the use of the machinery and equipment. It lays down working procedures to avoid foreseeable risks that cannot be eliminated or further reduced. As well as the operator, they should take account of possible risk to anyone else near enough to be affected by the way the work is done. The hazards and risks to be avoided should be identified during a risk assessment (*Chapter 4*).

Systems of work on their own should not be considered as safeguarding, but they do have an important role to play in complementing and backing-up the safeguarding arrangements that have been applied to machines. The agreed systems of work should be included in the skills training of operators and be part of the monitoring checks by supervisors.

All systems of work must be safe. Where the risks faced are low, instructions on the methods of work may be given verbally. However, where the risk faced is high and for the safety of the operator there must be no ambiguity in the understanding of the work method, the system of work should be in writing. Systems of work that are based on the results of a risk assessment or whose primary aim is to protect the operator are referred to as *safe systems of work*. In case of very high risks a *permit-to-work* system (*Appendix 6*) should be implemented and lay down specific steps and precautions to be taken before, during and after carrying out the work.

Preparation of systems of work is a management responsibility but local supervisors and operators should be involved since they have more detailed knowledge of the machinery and its operating risks. It will also ensure greater acceptability of the final system of work.

14.3 Protection from electric shock

14.3.1 Design considerations

Control panel equipment and machine wiring should be constructed to the requirements of IEC 60204 Part 1: which, in clause 6, deals with methods for reducing the risk of electric shock and burn injury. It applies to machines which operate at low voltages (LV) not exceeding 1000 V ac or 1500 V dc. Other parts of the standard may also be relevant.

Conductors energized at dangerous voltages should be protected to prevent any form of contact by unskilled persons with the minimum protection being complete enclosure or an effective barrier around the equipment when in normal production use. The protective measures should also prevent accidental contact, such as finger access to dangerous conductors, by skilled engineering employees while they are working on the equipment.

For equipment working in dry non-conductive environments, the control supply voltage should be as low as practicable to reduce the risk of shock. It should not exceed extra low voltage of 50 V ac or 120 V ripple free dc between conductors or between a conductor and the equipment enclosure which may be earthed or earth mass. (clause 9 of IEC 60204 Part 1).

In conductive environments, where, for example, equipment could be using water in its operation, there may be an increased risk of injury and lower voltages should be used to ensure safety.

14.3.2 Working on machine control panels

Dangerous conductors should not be manipulated while they are live. Modifications and continuity testing of circuits and components should only be carried out when circuits are isolated from the dangerous voltage.

During functional testing when machinery is not fully supervised, a level of protection equal to that for normal production should be maintained.

During maintenance and repair, when the normal protection may not be in place, local protection of conductors, such as temporary insulation, should be provided. Insulating mats can be used to prevent earth contact, but they will not be effective if contact can be made with the metal of the control panel which is likely to be earthed.

Test instruments used on high energy circuits should be of a type designed for such testing. They should have insulated probes to prevent finger contact and accidental short circuiting. Probes should incorporate fuses to protect against short circuit explosions within the instruments or between the test probe cables. For insulation testing the instrument output current should not exceed 5 mA. For high voltage insulation and flash testing where currents greater than 5 mA may be needed and several parts of the equipment may be energized simultaneously, special precautions should be taken. Class 1 mains powered test instruments such as oscilloscopes should not be used without the mains earth connection unless other means of shock protection are employed, such as double insulation (class 2).

Temporary power supplies to the circuits being tested and to the mains powered instrumentation should have, as supplementary protection, a residual current circuit breaker (RCCB) that detects leakage current of not more than 30 mA.

14.3.3 Portable equipment

Equipment that is hand held or regularly moved by hand should be supplied from a safe power source such as an extra low voltage source or conventional single or three phase mains supply protected by an RCCB. The extra low voltage supply can be from;

- double insulated isolating transformer with no earth reference at the output, or
- a generator with output insulation equivalent to a double insulated transformer with no earth reference, or
- a transformer or generator with a maximum 110 V output centre tapped to earth, or
- batteries or a dc generator.

The equipment and connecting cables should be inspected regularly for mechanical damage and electrically tested to ensure the level of safety is maintained.

14.4 Locking off

During maintenance and setting when it may be necessary to reach in or enter the machine, it is essential that inadvertent start-up is prevented. This can be achieved through 'locking off' the power source. This procedure involves locking, with a padlock or other similarly effective locking device, the operating handles of switches and valves in the safe position as shown in *Figure 14.1*. In general, electrical switches are locked in the open circuit condition, valves in feed pipelines are locked shut and valves in drain pipes are locked open.

Figure 14.1 Examples of locking off (a) an electrical switch and (b) a valve.

The person doing the work should apply the lock and keep the key on his person so that no one else can release the lock. Only the person who applies the lock should be permitted to remove it. Where two or more persons are involved in the work, a multi-hasp caliper, of the sort shown in *Figure 14.2*, should be used. This will ensure continuity of protection

Safe working with equipment   205

Figure 14.2 Multi-padlock hasp (or caliper).

until the last padlock has been removed. Where a gang or crew are involved, the charge hand or crew leader can apply a single lock on behalf of the gang or crew and is responsible for ensuring all members are clear and accounted for before removing the lock.

Padlocks should be marked so the 'owner' can be identified. Notices hung on switches and valves warning not to operate do not provide protection and should only be used to complement locking off.

An alternative method for preventing the start up of machines is by the withdrawal of fuses but this relies on an electrician being available and being responsible for ensuring everyone is clear before replacing the fuses. This method does not provide as high a level of protection as locking off but can be effective with small machines with local fuse boxes.

14.5 Ergonomics

Systems and methods of work are much more likely to be followed if the actions they require of the operator are comfortable, within his/her capacity, match his/her physical capabilities and do not put an undue stress or strain on him/her. Ergonomic principles should be followed in the design and contruction of the machinery, its controls, its guards and safeguarding arrangements. These should encompass the features outlined in *Chapter 7*. Controls and safeguards that are awkward to use can cause fatigue, stress and frustration with the likelihood that the operator will endeavour to by-pass or remove them giving rise to increased risks.

14.6 Anthropometrics

Anthropometrics is the study of measurements of the human body to determine averages within particular cultures. These data should be applied to the design of guards and safeguarding devices to ensure the measures taken provide effective protection for the greatest number of people. Account must be taken of national anthropometric data when designing guards especially for non-national customers. EN 547 – 3 gives body measurements averaged for Europeans. These data have been used to develop the standards for the safe distances of guards and barriers from points of danger for a range of work circumstances. Typical examples are discussed in *Chapter 8*.

The data in *Chapter 8* are particularly relevant when considering the physical dimensions of a guard in relation to its location from the point of hazard and the size of any gaps that might remain below, above, through or around the sides of guards.

14.7 Openings in guards

Openings are required in guards for a variety of reasons including feeding raw material, taking off finished product and enabling the operator to see the process. The size of opening will be determined by the reason for requiring it. Viewing can be through slots in a solid guard which should be small enough to prevent finger access or through gaps in a mesh but care must be taken to ensure the mesh is small enough to prevent hand access and far enough from the dangerous parts for them not to be reached by fingers.

Where the opening in the guard has to allow access for raw material or material holding jigs the design should be such that once the material or jig has been inserted into the processing position dangerous parts cannot be reached when the machine starts (*Chapter 5* and ENs 547–1, –2 and –3 and EN 811).

14.8 Operating instructions and manuals

Operating instructions and manuals should be provided by the supplier with all new machinery and should contain sufficient information concerning the machinery to:

- identify it consistent with the labels and marks on the machine;
- describe its intended use including any operating limits;
- give details of the component parts;
- enable assembly, commissioning and dismantling to be carried out safely;
- ensure safety in use, including identifying and giving warning of potential hazards associated with its use;
- allow setting and adjustment;
- enable it to be maintained and repaired.

Accompanying the machine should be documentation specifying the operating limits of the machine. Manufacturers of machinery for use in the EU must keep a technical file containing details of the design and of the standards to which the machinery has been designed and constructed.

If the user wishes to vary the operating procedure, or has modified the machine to meet particular production needs, he must carry out a risk assessment to ensure that the new method of work is at least as safe as that given in the maker's operating instructions.

14.9 Labels on equipment

Attached to all machines should be permanent labels containing the following information:

- name and address of manufacturer;
- designation of type or series of machine;
- serial number;
- year of manufacture;
- any machine to be marketed and used in the EU, the CE mark.

Where material used in the process presents a particular hazard, warning notices should be attached to the machine identifying the material concerned, the hazards it presents and the precautions to be taken. This can be by the use of standard pictograms.

14.10 Supervision

Effective control and safe use of machinery can be improved through the presence of adequate and competent supervision. The supervisors must be knowledgeable in the work being carried out and experienced in the use of the particular machinery. They should be vested with sufficient authority to enable them to initiate safety actions up to a certain level of expenditure and to take any disciplinary action necessary to ensure adherence to the agreed working methods and procedures.

14.11 Use of jigs and fixtures

In certain specific operations where the use of a guard either prevents the work being done or increases the hazard, a jig or fixture may be used to hold the work while keeping the operator's hands away from the dangerous part of the machine. Typical of such operations include the use of rotary moulding machines and routers in carpentry and cabinet making.

14.12 Safety clothing

Loose clothing worn in the vicinity of machinery can be caught by moving parts and result in the wearer being drawn into the machine. Close fitting protective clothing without loose cuffs, pockets, etc. should be provided and worn. Where a machine can emit potentially hazardous materials, special personal protective equipment should be issued, for example spark proof clothing for welders and fettlers, chemical (green) suits for work with hazardous chemicals, etc.

14.13 Stored energy

Where stored energy is used as part of the process or occurs as part of the machine function, before maintenance, adjustment or repair work is commenced the source of the energy must be isolated from the machine and any stored energy remaining in the machine dissipated. Stored energy can be as:

- compressed air,
- steam,
- hydraulic circuits with accumulators,
- compressed springs,
- heavy components or counter-weights in a raised position,
- electrical capacitors.

Compressed air and steam should be vented to atmosphere, hydraulic systems discharged to sump, the compression of springs released and heavy components retained in position by mechanical scotches, secured props or lowered to a suitable support or the floor.

Care must also be taken to ensure that programmable controllers are isolated from the machine controls or that the software has been designed to prevent machine movement after isolation. Special precautions must be taken with robots in teach mode when operators have to work in close proximity to live machines.

14.14 Signs and signals

Where signs are attached to the machine to indicate either hazards or precautions to be taken, they should be of internationally recognized pictograms and may be backed-up by written legend in the language of the country of use.

Where it is necessary to use signals to communicate between two or more operators, the signals used should be clear, unambiguous and understood by all the parties involved. In the use of cranes and lifting equipment only one individual, the banksman or slinger, should give signals (instructions) to the crane operator. Similarly in marshalling yards, only the shunter should give signalled instructions to the locomotive driver. The signals used should be to an international standard such as are illustrated in ENs 457, 842 and 981.

Chapter 15

Plant layout and the working environment

15.1 Introduction

The use of plant and machinery is not confined to enclosed workshops with heating, lighting and the other services accepted as necessary parts of the working conditions. Some process plants, such as petro-chemical plants, are too large to house in a building. Other workplaces using machinery are in the open with minimal weather protection while mobile equipment tends to be used completely in the open. Whether in a building or the open, the proper layout of machinery has an important role in reducing the risks to which the operators are exposed to a minimum.

Similarly, the environment of a workplace can have a material effect upon the efficiency and life of machinery. A hostile or corrosive atmosphere can damage parts of the machine, temperature can affect the accuracy in tool room work while the general cleanliness can influence the manner in which the machines are used and hence affect productivity.

Machines should only be used in environments which are compatible with the materials of the machine and the materials used in the process. Where the environment in which a machine has to work is likely to affect adversely machine components, such as seals, bearings, lead screws, piston rods, etc., they must be made of, coated with or protected by materials that will resist the corrosive effects. The machine designer must ensure that any effects from the environment do not increase the risks to the operator or others from the use of the machine.

15.2 Space

Adequate space should be left around all machinery to ensure it can be operated safely, that all necessary setting, adjustment and maintenance can be carried out, that raw material can be supplied to and finished goods removed from it. Operators should have safe access to all parts it is necessary to approach. Where power trucks or mobile cranes are employed the gangways need to be of suitable width to permit any manoeuvres the vehicles have to make.

15.3 Buildings

Buildings need to be of suitable construction to protect the machinery they house, primarily rain-tight with other facilities to provide levels of weather protection and environmental conditions suitable for the plant housed and for operator comfort.

The structure needs to be able to accommodate any fixed lifting equipment and the various runs of services.

15.4 Services

Provision must be made for the runs of services required by the various machines including electrical power, air supplies, water, ventilation, lighting, drains, etc.

The installed electrical capacity must be adequate to meet the demands of the machines without overloading the supply circuits. Each machine should be provided with its own isolator [IEC 60204–1] complete with locking-off facility (*section 14.3*).

The line supplying air to each machine should be of adequate size for the machine's needs (*section 13.2.1*) and, where required, be provided with quick release couplings (*section 13.2.6*) for pneumatic hand tools and hand-held air jets for blowing down or cleaning. The air supply to each machine should be provided with a stop valve to enable the machine to be isolated from the supply.

A water supply should be laid on if the machine process requires it when provision should also be made for suitable drains to remove waste water.

15.5 Ventilation

Where a machine is working in or creates a hostile atmosphere giving rise to the risk of damage to machine parts or product, suitable ventilation may be needed. On machines that generate more heat than can be absorbed within the workplace atmosphere, an extraction system should be provided, and consideration may need to be given to the provision of local ventilation for the operator. Heat exchangers should be considered so that the extracted heat can be utilized for space heating, water heating, etc.

The environment in which machine operators have to work should not put their health at risk. Where noxious fumes, dusts, contamination, high humidity and temperatures are likely to occur suitable extraction and/or fresh air ventilation should be provided. These facilities can be general for the whole work area or local to the specific area in which the operator works, such as the control rooms of large automated plant.

15.6 Lighting

The general level of lighting to be provided in the work place will depend on the type of work being carried out but an adequate level of lighting should be provided at the operating positions of machinery. Levels of lighting for typical working environments are:

Location and task	Standard service illuminance (lx)
Storage areas, plant rooms, entrance halls, etc.	150–200
Rough machinery and assembly, canteens, control rooms, wood machinery, cold strip mills, weaving and spinning, etc.	300–400
Routine office work, medium machinery and assembly, etc.	500
Workplaces in which vdts are used regularly	300–500
Demanding work such as drawing offices, inspection of medium machinery, etc.	750
Fine work requiring colour discrimination, textile processing, fine machining and assembly	1000
Very fine work such as hand engraving and inspection of fine machinery and assemblies	1500

In the layout and positioning of luminaires care must be taken to ensure there are no positions of sharp light and dark contrast and that the operators are not subject to dazzle or glare. Wherever possible natural daylight should be used but in sunny conditions some shading may be necessary.

Supplementary lighting may be necessary at the control panel to ensure accurate reading of instruments and control positions. An ability to adjust local lamp positions can be helpful. Where viewing of work in progress is necessary additional lighting within the machine may be needed. The type of lighting can suit local circumstances but if fluorescent or other non-incandescent lighting is used care must be taken to prevent a stroboscopic effect with rotating machine parts. This can be achieved by utilizing twin tubes supplied from electrical supply sources that are 90° out of phase. When positioning local lamps care must be taken to ensure they do not create glare, local areas of deep shadow or dazzle the operator.

15.7 Temperature

Normally the temperature of the environment in which machinery operates is dictated by operator comfort. However, there may be circumstances in which machinery has to operate at elevated temperatures (steel and paper mills) or depressed temperatures (chiller stores). The temperature at which the machinery is to operate must be allowed for in its design.

Generally, the temperature of a work area will depend on the type of work being carried out. For sedentary and close work a temperature around 20–22°C may be required while for more physical work a temperature of 13–16°C may be adequate. In special applications, such as tool rooms where accuracy is of paramount importance, the environment will need to be strictly controlled both for temperature and humidity.

15.8 Machine layout

In laying out or positioning machinery regard must be had for the manner in which it is to be operated. Consideration should be given to what the operator needs to do in the day-to-day operations, setting, adjustment, servicing and maintenance and whether materials and goods will be handled manually, by conveyor or by powered truck.

While guarding and protective devices provide protection at the operator's position and those parts adjacent to walkways, the prevention of access to dangerous parts at the back of the machine must also be considered.

Machines should be positioned so that when components or parts are at the extremity of their movement they do not create trapping points with other parts of the machine or the building structure (*section 3.3.2.c*)

15.9 Noise

Noise has long been recognized as a health hazard. Ideally noise levels should be reduced to the minimum possible to ensure that operator's hearing is not at risk of damage. An average daily noise level dose of 85 dBA or below is not likely to present a risk. Above this level hearing is put at risk and personal hearing protection should be worn. Above 90 dBA personal hearing protection must be worn. The aim should be to reduce noise emissions rather than provide personal protection.

In enclosed workplaces, the contribution made by each machine to the general noise level should be noted. This is particularly relevant if the background noise level is high and additional machines are to be installed. Where additional machines are to be installed in a work area an assessment should be made of the likely increase in the overall noise level they may cause.

Designers should aim to produce machinery that emits a noise level below 85 dBA. Where this level is exceeded the user should be warned. At and above 85 dBA noise level in the work area consideration should be given to noise reduction measures, both on the machines and to the building. Noise reduction techniques can include bracing vibrating panels, lining panels with sound absorbing materials, using sound absorbing screens around particularly noisy machines or throughout the noisy area to prevent transmission of sound. Also the use of plastic gears and plastic bushes in linkages but care must be taken to ensure the material used will give an effective and economic performance. Effective maintenance also contributes to keeping down noise levels.

15.10 Vibrations

Machines whose operations generate vibrations should be arranged so that these effects are not transmitted to the operators or to adjacent machines or the building. Such machines should have special anti-vibration mounts. Where pipes and ducts are connected to machinery that vibrates, flexible connections should be employed. Low frequency and air transmitted vibrations present a particular risk to body organs and special precautions should be taken to reduce these risks to a minimum. Hand-held tools that vibrate can cause vibration white finger and should be fitted with vibration proof handles.

15.11 Materials handling

The means for delivering raw materials and removing finished goods will be dictated by the product. Where materials and finished goods are manually handled, loads should be within the lifting capacity of the handlers who should be trained in safe lifting techniques. If mechanical handling, such as conveyors, are used they must be fully guarded (*section 3.3.1.e* and *3.3.3.c*). If lift trucks are used, suitable width gangways to allow proper manoeuverability must be provided and the truck operators fully trained and certificated (licensed).

Storage of materials – raw and finished – in the machine vicinity should not interfere with its operation, setting or adjustment nor with access ways. Neither raw materials nor finished goods should be stored in front of switchgear, emergency exits or in accessways.

15.12 Maintenance

Adequate facilities to carry out all the necessary maintenance should be provided. This should include space to enable components to be removed from the machine, lifting facilities for heavy items and power supply points for portable tools – both electric and pneumatic.

Safe systems of work should be prepared for all maintenance work and should include arrangements for the isolation of power supplies – locking off (*section 14.4*) – and the implementation of permits-to-work (*Appendix 6*) where the work involves high risks.

15.13 Waste

In many production processes there are spillages and breakages that result in contamination of the machine and the surrounding work area. Such contamination should be cleaned up as soon as possible and any debris removed since it can interfere with the machine performance and can cause a slipping and tripping hazard to the operators. Special absorbent materials should be used for liquid spillages.

Surplus materials and waste from the process should be collected in suitable containers or bins and disposed of in an environmentally friendly way. Certain types of waste can have a commercial value. The disposal of hazardous waste must follow the laid down procedures for such materials.

15.14 Access

Clear access must be provided to all positions from which it is necessary to operate the machine. This access should not be impeded by adjacent machines, raw material or finished stock. Floor markings should be used to indicate the areas to be kept clear.

Where a raised platform is needed to reach the machine controls, or other parts necessary for the operation of the machine, suitable access steps with handrail should be provided. Vertical ladders should have safety hoops. If there is a risk of the operator falling from the operating platform, it should be provided with a handrail. Operating platforms should be level, provide adequate floor area for the safe operation of the controls and be strong enough to support the operating loads, especially if raw material or finished goods are stacked on them. The surfaces of platforms and steps should be slip resistant. Similar facilities should be provided where access is required for setting, adjusting, cleaning and maintenance.

Provision should be made for safe access to the areas below a machine for cleaning purposes. This is particularly important with conveyors, transfer machines, robots and machining centres.

15.15 Lubrication

Wherever possible, lubrication should be carried out from outside the guards of a machine. Automatic lubrication systems and remote manual systems for both oil and grease can achieve this but should be provided with failure (empty) warning devices. If it is necessary to enter the machine to lubricate it, access should be through interlocking guards or when the machine is shut down. Care must be taken in the application of lubricants since excess can drip on the floor creating a slipping hazard. Surplus lubricant around a bearing will attract fluff and dust, and in extreme cases, create a fire hazard. Bearings running in a hostile atmosphere should be of the sealed type.

If the lubricant of a bearing can be seen it is not doing its job.

15.16 Dust and fumes

Dusts and fumes should not be allowed to accumulate in the vicinity of machines. Where dusts and fumes are generated by the process or by the plant suitable exhaust ventilation or extraction should be provided. Certain processes such as food preparation, pharmaceuticals, nuclear, microprocessor manufacture, etc., demand very high standards of cleanliness and freedom from foreign matter requiring special air conditioning plant.

15.17 Floors and foundations

The floors of workplaces should be suitable for the equipment and machinery to be installed and for the vehicles that use it. The foundations to which machines are attached should be suitable for the size and weight of machine. Areas in which fork lift trucks operate should have floors that are smooth, level, well drained and durable.

15.18 Hygiene

In areas handling food, pharmaceuticals and other products where hygiene is critical the design and layout of plant should allow effective cleaning to be carried out. Details of requirements and techniques that can be employed are given in ISO 14159.

15.19 Notices and signs

Any signs, notices or warnings necessary for the safety of the machine should be permanent, on durable material, clearly displayed and readily understandable. They should be of the standard symbolic or pictogram type to EN 842 with legend if required.

Part V
Appendices

Appendix 1
Published standards

There are three major standard-making bodies, each identified by prefix letters to the standard number:

European	EN
International Electrotechnical Commission	IEC
International Standards Organisation	ISO

These three bodies now work closely together with the result that there is a growing number of standards that have been adopted as both European and international standards. These can be identified by double prefix letters, i.e. EN.IEC and EN.ISO. This list is split into three parts corresponding to the particular originating organization and only its set of prefix letters used.

The most prolific standard-making body is the European (EN) – through the Comité Européen de Normalisation (CEN) and Comité Européen de Normalisation Electrotechnique (CENELEC). When an EN standard is adopted internationally it is given a five figure number. Where this has occurred, or is occurring, the new international number is given in square brackets [] at the end of the title. Similarly there is a cross-reference from international standards to the equivalent EN standard. A number of the standards listed have not yet been finalized and adopted. These have been indicated by the prefix 'pr'.

This list of standards is restricted to those standards that lay down principles for achieving safety and that have general application. It is far from exhaustive but does cover those most likely to be needed by the designer and engineer. There are many other standards that are 'machine specific' and relate particularly to specific types of machinery.

Copies of these standards can be purchased from the national standard making organization, but they can be expensive. However, many reference libraries have micro-fiche or electronic copies that can be consulted.

European standards

EN 292 Part 1	Safety of machinery – Basic concepts and general principle for design – Part 1: Basic terminology and methodology. [pr ISO 12100-1]
EN 292 Part 2	Safety of machinery – Basic concepts and general principles for design – Part 2: Technical principles and specifications. [pr ISO 12100-2]
EN 294	Safety of machinery – Safety distances to prevent danger zones being reached by the upper limbs. [ISO 13852]
EN 349	Safety of machinery – Minimum gaps to avoid crushing of parts of the human body. [ISO 13854]
prEN 414	Safety of machinery – Rules for drafting and presentation of safety standards.
EN 418	Safety of machinery – Emergency stop equipment, functional aspects – Principles for design. [ISO 13850]
EN 457	Safety of machinery – Auditory danger signals – General requirements, design and testing (see ISO 7731 latest issue).
EN 547-1	Safety of machinery – Human body measurements – Part 1: Principles for determining the dimensions required for access openings for the whole body access into machinery.
EN 547–2	Safety of machinery – Human body dimensions – Part 2: Principles for determining the dimensions required for access openings.
EN 547–3	Safety of machinery – Human body measurements – Part 3: Anthropometric data.
EN 563	Safety of machinery – Temperatures of touchable surfaces – Ergonomic data to establish temperature limit values for hot surfaces.
EN 574	Safety of machinery – Two-hand control devices – Functional aspects – Principles for design. [ISO 13851]
EN 614–1	Safety of machinery – Ergonomic design principles – Part 1: Terminology and general principles.
EN 626–1	Safety of machinery – Reduction of risks to health from hazardous substances emitted by machinery – Part 1: Principles and specifications for machinery manufacturers. [ISO 14123–1]
EN 626–2	Safety of machinery – Reduction of risks to health from hazardous substances emitted by machinery – Part 2: Methodology leading to verification procedures. [ISO 14123-2]
EN 811	Safety of machinery – Safety distances to prevent danger zones being reached by the lower limbs. [ISO 13853]
EN 842	Safety of machinery – Visual danger signals – General requirements, design and testing.
EN 894–1	Safety of machinery – Ergonomic requirements for the design of displays and control actuators – Part 1: General

	principles for human interactions with displays and control actuators. [ISO 9355–1]
EN894–2	Safety of machinery – Ergonomic requirements for the design of displays and control actuators – Part 2: Displays. [ISO 9355–2]
EN 953	Safety of machinery – Guards – General requirements for the design and construction of fixed and movable guards. [ISO 14120]
EN 954-1	Safety of machinery – Safety-related parts of control systems – Part 1: General principles for design [see ISO 13849–1].
prEN 954-2	Safety of machinery – Safety related parts of control systems – Party 2: Validation. [ISO 13849–2]
EN 981	Safety of machinery – System of auditory and visual danger and information signals.
EN 982	Safety of machinery – Safety requirements for fluid power systems and their components – Hydraulics.
EN 983	Safety of machinery – Safety requirements for fluid power systems and their components – Pneumatics.
EN 999	Safety of machinery – The positioning of protective equipment in respect of approach speeds of parts of the human body. [ISO 13855]
EN 1012-1	Compressors and vacuum pumps – Safety requirements. Part 1 Compressors.
EN 1012-2	Compressors and vacuum pumps – Safety requirements. Part 2 Vacuum pumps.
EN 1037	Safety of machinery – Prevention of unexpected start-up. [ISO 14118]
EN 1050	Safety of machinery – Principles for risk assessment. [ISO 14121]
EN 1088	Safety of machinery – Interlocking devices associated with guards – Principles for design and selection. [ISO 14119]
EN 1093-1	Safety of machinery – Evaluation of the emission of airborne hazardous substances – Part 1: Selection of test methods.
EN 1093-3	Safety of machinery – Evaluation of the emission of airborne hazardous substances – Part 3: Emission rate of a specified pollutant – Bench test method using the real pollutant.
EN 1093-4	Safety of machinery – Evaluation of the emission of airborne hazardous substances – Part 4: Capture efficiency of an exhaust system – Tracer method.
EN 1093-6	Safety of machinery – Evaluation of the emission of airborne hazardous substances – Part 6: Separation efficiency by mass, unducted outlets.
EN 1093-7	Safety of machinery – Evaluation of the emission of airborne hazardous substances – Part 7: Separation efficiency by mass, ducted outlet.

EN 1093-8	Safety of machinery – Evaluation of the emission of airborne hazardous substances – Part 8: Pollutant concentration parameter, test bench method.
EN 1093-9	Safety of machinery – Evaluation of the emission of airborne hazardous substances – Part 9: Pollutant concentration parameter, room method.
prEN 1127-1	Safety of machinery – Explosive atmospheres – Explosion prevention and protection – Part 1: Basic concepts and methodology.
prEN 1175-1	Safety of machinery – Industrial trucks – Part 1: Electrical requirements for battery electrical trucks.
prEN 1175-2	Safety of machinery – Industrial trucks – Part 2: Electrical requirements for internal combustion engined power trucks.
prEN 1175-3	Safety of machinery – Industrial trucks – Part 3: Electrical requirements for electric power transmission systems of internal combustion powered trucks.
EN 1299	Mechanical vibration and shock – Vibration isolation of machines – Information for the application of source isolation.
EN 1760-1	Safety of machinery – Pressure sensitive protective devices – Part 1. General principles for the design and testing of pressure sensitive mats and pressure sensitive floors. [ISO 13856-1]
prEN 1760-2	Safety of machinery – Pressure sensitive protection devices – Part 2: General principles for the design and testing of pressure sensitive edges and pressure sensitive bars. [ISO 13856-2]
prEN 12198-1	Safety of machinery – Assessment and reduction of risks arising from radiation emitted by machinery – Part 1: General principles.
EN 12626	Safety of machinery – Laser processing machines – Safety requirements [see ISO 11553 latest issue]
prEN 50081-2	Electromagnetic compatibility – Generic emission standard – Generic standard class – Part 2: Industrial environment.
prEN 50099 -1	Safety of machinery – Indication, marking and actuation – Part 1: Requirements for visual, auditory and tactile signals [see EN 61310-1]
prEN 50099-2	Safety of machinery – Indication, marking and actuation – Part 2: Requirements for marking.
prEN 50179	Power installations exceeding 1 kV ac.
EN 60204-1	Safety of machinery – Electrical equipment of machines – Part 1: General requirements [IEC 60204-1].
EN 61310-1	Safety of machinery – Indication, marking and actuation – Part 1: Requirements for visual, auditory and tactile signals. [IEC 61310–1].
EN 61310-2	Safety of amchinery – Indication, marking and actuation – Part 2: Requirements for marking [IEC 61310-2].

EN 61310-3   Safety of machinery – Indication, marking and actuation – Part 3: Requirements for the location and operation of actuators [IEC 61310-3].
EN 61496-1   Safety of machinery – Electro-sensitive protective equipment – Part 1: General requirements and tests. [IEC 61496-1]
EN 61496-3   Safety of machinery – Electro-sensitive protective equipment – Part 3: Particular requirements for equipment using active opto-electronic devices responsive to diffuse reflections (AOP DORs). [IEC 61496-3]
prEN 61496-4 Safety of machinery – Electro-sensitive protective equipment – Part 4: Particular requirements for equipment using passive infra-red protective devices.
R044-001   Safety of machinery – Guidance and recommendations for the avoidance of hazards due to static electricity.

International mechanical standards

ISO 7250   Basic human body measurements for technological design.
ISO 9000   Quality management and quality assurance standards
             Part 1: Guidelines for selection and use
             Part 2: Generic guidelines for the application of ISO 9001, ISO 9002 and ISO 9003
             Part 3: Guidelines for the application of ISO 9001 to the development, supply, installation and maintenance of computer software.
ISO 9001   Quality systems. Model for quality assurance in design, development, production, installation and servicing.
ISO 9002   Quality systems. Model for quality assurance in production, installation and servicing.
ISO 9003   Quality systems. Model for quality assurance in final inspection and test.
ISO 9004   Quality management and quality assurance standards. Part 1: Guidelines
ISO 11161   Industrial automation systems – Safety of integrated manufacturing systems – Basic requirements.
ISO 12100-1 Safety of machinery – Basic concepts and general principles for design – Part 1: Basic terminology and methodology. [EN 292-1]
ISO 12100-2 Safety of machinery – Basic concepts and general principles for design – Part 2: Technical principles. [EN 292-2]
ISO 13848   Safety of machinery – Terminology.
ISO 13849   Safety of machinery – Safety-related parts of control systems – Part 1: General principles for design. [EN 954-1]
ISO 13849-2 Safety of machinery – Safety-related parts of control systems – Part 2: Validation, testing and fault lists. [EN 954-2]

ISO/TR 13849	Safety of machinery – Safety-related parts of control systems – Part 1: General principles for design
ISO 13850	Safety of machinery – Emergency stop – Principles for design. [EN 418]
ISO 13851	Safety of machinery – Two-hand control devices – Functional aspects and design principles. [EN 574]
ISO 13852	Safety of machinery – Safety distances to prevent danger zones being reached by the upper limbs. [EN 294]
ISO 13853	Safety of machines – Safety distances to prevent danger zones being reached by the lower limbs. [EN 811]
ISO 13854	Safety of machinery – Minimum gaps to avoid crushing of parts of the human body. [EN 349]
ISO 13855	Safety of machinery – Positioning of protective equipment with respect to the approach speeds of parts of the human body. [EN 999]
ISO 13856-1	Safety of machinery – Pressure-sensitive protective devices – Part 1: General principles for design and testing of pressure-sensitive mats and pressure-sensitive floors. [EN1760-1]
ISO 13856-2	Safety of machinery – Pressure-sensitive protective devices – Part 2: General principles for the design and testing of pressure-sensitive edges and pressure-sensitive bars. [EN 1760-2]
ISO 13856-3	Safety of machinery – Pressure-sensitive protective devices – Part 3: General principles for the design and testing of bumpers. [EN 1760-3]
ISO 14001	Environmental management systems. Specification with guidelines for use.
ISO 14004	Environmental management systems. General guidelines on principles, systems and supporting techniques.
ISO 14118	Safety of machinery – Prevention of unexpected start-up. [EN 1037]
ISO 14119	Safety of machinery – Interlocking devices with and without guard locking – General principles and precision for design. [EN 1088]
ISO 14120	Safety of machinery – Guards – General requirements for the design and construction of fixed and movable guards. [EN 953]
ISO 14121	Safety of machinery – Principles of risk assessment. [EN 1050]
ISO 14122-1	Safety of machinery – Permanent means of access to machines and industrial plants – Part 1: Choice of a fixed means of access between two levels.
ISO 14122-2	Safety of machinery – Permanent means of access to machinery – Part 2: Working platforms and walkways.
ISO 14122-3	Safety of machinery – Permanent means of access to machinery – Part 3: Stairways, stepladders and guardrails.
ISO 14122-4	Safety of machinery – Permanent means of access to machinery – Part 4: Fixed ladders.

ISO 14123-1   Safety of machinery – Reduction of risks to health from hazardous substances emitted by machinery – Part 1: Principles and specifications for machinery manufacturers. [EN 626 -1]
ISO 14123-2   Safety of machinery – Reductions of risks to health from hazardous substances being emitted by machinery – Part 2: Methodology leading to verification procedures. [EN 626-2]
pr ISO 14159   Safety of machinery – Hygiene requirements for the design of machinery.
ISO 13849-1   Safety of machinery – Safety related parts of control systems – Part 1: General principles for design.
ISO 18569   Safety of machinery – Guideline to the understanding and use of safety of machinery standards.

International electrical standards

IEC 417   Graphic symbols for use on equipment – Harmonisation of texts for IEC 417.
IEC 1081-11   Preparation used in electrotechnology – Part 11: Preparation of instructions.
IEC 1346-2   Industrial systems, installations and equipment and industrial products – structuring principles and reference designations – Part 2: Classification of objects and codes for classes.
IEC 60204-1   Safety of machinery – Electrical equipment of machines – Part 1: General requirements.
IEC 60204-11   Safety of machinery – Electrical equipment of machines – Part 11: General requirements for voltages above 1 000V ac or 1 500 V dc and not exceeding 36kV.
IEC 60204-31   Safety of machinery – Electrical equipment of machines – Part 31: Particular safety and EMC requirements for sewing machines, units and systems.
IEC 60204-32   Safety of machinery – Electrical equipment for machines – Part 32: Requirements for hoisting machines.
IEC 60364   Electrical installations of buildings – 6 parts
IEC 60446   Basic safety principles for man machine interface, marking and identification – Identification of conductors by colour or numerals.
IEC 60479-1   Effects of current passing through the human body – Part 1: General aspects
IEC 60479-2   Effects of current passing through the human body – Part 2: Special effects
IEC 60529   Degrees of protection provided by enclosures (IP code).
IEC 61032   Protection of persons and equipment by enclosures – Probes for verification.
IEC 61069-7   Industrial process measurement and control – Evaluation of system properties for the purpose of system assessment – Part 7: Assessment of system safety.

IEC 61140	Protection against electric shock – Common aspects for installation and equipment.
IEC 61310-1	Safety of machinery – Indication, marking and actuation – Part 1: Requirements for visual, auditory and tactile signals [EN 61310-1].
IEC 61310-2	Safety of machinery – Indication, marking and actuation – Part 2: Requirements for marking [EN 61310-2].
IEC 61310-3	Safety of machinery – Indication, marking and actuation – Part 3: Requirements for the location and operation of actuators [EN 61310-3].
IEC 61340-1	Guide to the principles of electrostatic phenomena.
IEC 61340-2-3	Methods of test for determining the resistance and resistivity of solid planar materials used to avoid electrostatic charging.
IEC 61491	Electrical equipment for industrial machines – Serial data link for real-time communications between controls and drives.
IEC 61496-1	Safety of machinery – Electro-sensitive protective equipment – Part 1: General requirements and tests [EN 61496-1].
IEC 61496-2	Safety of machinery – Electro-sensitive protective equipment – Part 2: Particular requirements for equipment using active opto-electronic protective devices (AOPDs).
IEC 61496-3	Safety of machinery – Electro-sensitive protective equipment – Part 3: Particular requirements for equipment using active opto-electronic devices responsive to diffuse reflections (AOPDDRs) [EN 61496-3].
IEC 61508	Safety of machinery – Functional safety of electrical, electronic and programmable electronic safety related systems.
IEC 62046	Safety of machinery – Application of personnel sensing equipment to machinery.
IEC 62061	Safety of machinery – Functional safety – Electrical, electronic and programmable electronic control systems.

Appendix 2
Glossary of terms

Adjustable guard A fixed guard that has adjustable elements that can be arranged to accommodate the profile of a particular workpiece. Once positioned, the elements are locked in position.
Anthropometrics The study of the physical proportions of humans with the view to determining average dimensions for different sexes, ages, and ethnic groups.
Back-up safety function That function that complements safety-critical and safety-related functions and whose failure does not result in immediate hazard although it may reduce the inherent level of safety.
Braking The act of bringing to a halt. It can be achieved by friction devices or by electrical means.
Common cause failure Where the failure of two or more safety devices can be attributed to the same root cause or event and where the failures are not related sequential events.
Common mode failure In which the failure by a number of devices to perform a safety function has as its cause a common misfunction of a component part or system even though the causes may differ.
Control interlocking An arrangement whereby the interlocking device interacts with the controls to bring a machine to a stop or safe condition.
Crawl control A control which when actuated causes the machine to run at a predetermined slow (crawl) speed.
Cross connection faults Faults in an electrical circuit in which electrical leakage occurs between adjacent connections.
Danger The state of being at risk and in a position where personal harm or injury is foreseeable.
Defeatability Any weakness in the safeguarding arrangements for a machine that allows them to be negated or by-passed hence reducing the level of protection provided.
Diversity The use of different media in associated safety circuits to reduce the likelihood of common cause failure.
Duplication The inclusion in a safety circuit of duplicate safety devices to reduce the likelihood of a common mode failure occurring.
Ergonomics The study of the relationship between the human and the equipment being operated with a view to achieving its most efficient, effective and satisfactory use.
Fail to danger A circumstance under which the failure of any part of the machine, its control equipment or power supply leaves the machine in an unsafe condition.

Fail to safety An ideal concept where on failure of any part of the machine, its control equipment or power supply the machine remains in a safe condition. (Note: this concept is unlikely to be realized in practice – a better phrase may be **not fail to danger**.)
Failure The termination of the ability of an item to perform a required function.
Failure cause The circumstance, whether identified at the design stage or experienced in use, that results in a failure.
Failure criteria The conditions under which an item is most likely to fail or the conditions to be considered and avoided to ensure an item does not fail.
Failure mechanism The process, whether physical, chemical, environmental or other, that results in the failure of an item.
Failure mode The way or manner in which a failure occurs.
Failure to danger Any failure of a machine or any part of it that creates a hazardous situation.
Fast break switch A switch in which the close and open action is controlled by a spring that is independent of the switch actuator movement.
Fault A condition in an item that prevents it from performing its intended function.
Fence A barrier placed around a dangerous machine or machinery to prevent access to parts likely to cause damage or harm.
Fieldbus An electronic link carrying transmitted information over a limited area or field.
Fixed guard A guard fixed in position by means that require the use of a tool to release it.
Functional safety That aspect of safety reliant upon the proper functioning of the equipment.
Guard A physical barrier, whether fixed, movable or removable, that prevents access to a dangerous part, area or zone.
Hazard Something with the potential to cause damage or harm.
Hazardous machine function Any function or operation of a machine that creates a hazard for the operator or others.
Hold-to-run control A control device which when manually actuated causes the machine to run at a predetermined reduced speed. Release of the control automatically brings the machine to rest.
Integrity The ability of the machine and any components, safety or otherwise, to perform their functions without failure for a predetermined period of time or number of functions.
Interlock A safety device actuated by a guard and connected to the controls of or the power supply to a machine so as to cause the dangerous part of the machine to cease movement or the machine to revert to a safe condition.
Interlocking device A device, whether of mechanical, electrical, pneumatic, hydraulic or other medium, that is actuated by a machine guard and interacts with the machine controls to cause the machine to stop or revert to a safe condition.
Interlocking guard A movable guard incorporating an interlocking device.

Luminaire A general term covering all the apparatus required to produce a lighting effect, i.e. the complete light fitting.

Machine An apparatus or equipment, whether manual or power operated, for changing or manipulating materials and used in the manufacture of a product. It comprises a series of fixed or moving parts that perform the various functions to achieve the required end result.

Machinery An assembly of linked parts or components, one or more of which moves, with the object of processing, treating, moving or packaging a material. It may be part of a larger machine or could comprise a series of linked machines performing a particular operation.

Maintainability The extent to which the design and layout of the machine allows it to be serviced and maintained so it can continue performing to the level of the maker's specification.

Mechanical restraint device A mechanical component which when inserted into the appropriate machine part prevents further movement of the machine or part (sometimes referred to as scotch, wedge, strut, or prop, etc.).

Monitoring The process of checking on the ability of a safety function to continue performing satisfactorily. Monitoring can be:

- continuous, with repeated automatic carrying out of checks at regular predetermined time intervals;
- discontinuous, with checks initiated by an event or occurrence such as the completion of an operational cycle, break in production flow, shift change, etc., or
- random checks, initiated by the operator or another at such times as he/she thinks fit.

In each case if a failure is detected, a safety measure is initiated either to maintain the required level of safety for continuing operations or to return the machine to a safe condition.

Movable guard A guard where its attachment to a machine is such that it can be moved but not removed from the machine.

Movement limiting device A part of the machine controls which when actuated permits the machine product to move a limited distance (~75mm) or for the machine to run for a limited time.

Muting A condition whereby the safety device or parts of it can be isolated to enable particular operations to be carried out in or near an area of danger. Normally only used with electronic safety devices or controls and restricted to specific parts of a machine's operations where the increase in risk is minimal.

Not fail to danger A condition where failure of any part of the machine, its control equipment, its safety devices or power supply does not increase the risks faced by the operator or those working in the vicinity.

Pipeline A discrete run or length of pipe in an overall pipework system such as the main pipe carrying the service to the location where it is to be used.

Pipework A general term referring to various miscellaneous pipes that make up an overall piping system.

Power interlocking An interlocking arrangement whereby actuation of the interlocking device causes the power supply to the machine to be interrupted.

Pressure reducing valve (PRV) A valve provided to reduce the pressure in a service pipe from the main supply pressure to a lower pressure to meet a particular operational need.

Pressure relief valve see *Safety valve*

Probability The possibility or likelihood of an event occurring. It is usually expressed as a ratio or percentage.

Redundancy The installation in parallel (duplication) of similar equipment so that on a failure of one set the other will maintain the desired level of protection. The use of duplicate equipment reduces, but does not entirely eliminate, the risks from common mode failure.

Reliability The ability of the machine, component or device to continue performing its function over its expected working life.

Removable guard A guard that can be completely removed from the machine or equipment to allow access for setting, adjustment, maintenance, etc. When replaced it should be secured by means requiring a tool to release it or be held in position by retaining interlocking devices.

Residual hazard The hazard, identified by a risk assessment, that remains after all attempts to eliminate or reduce it have proved unsuccessful.

Residual risk The risk that remains after protective measures, identified by the risk assessment, have been applied.

Risk A combination of the likelihood of the injury occurring and the likely extent of the harm or injury.

Risk assessment A procedure for estimating the risks faced by operators and others from the hazards remaining after all attempts to eliminate them by design have been exhausted. This phrase is sometimes used in a general sense to include the hazard identification and elimination process.

Risk reduction A procedure undertaken to reduce to a minimum the risks from the residual hazards of a machine. It is based on the findings of a risk assessment. For the makers of new machines, risk reduction should be carried out at every manufacturing stage from initial design to final production. For users of existing machines, risk reduction will form part of the regular risk assessments but may also be initiated by an event such as an accident, near miss, machine failure or breakdown, etc.

Safeguard A guard, device or combination of devices, controls and systems aimed at protecting a person from danger.

Safeguarding device A guard or other device that provides protection to an operator from the hazards of a machine.

Safe system of work A considered method of working that takes proper account of the potential hazards faced and lays down methods of work to avoid or reduce to a minimum the possibility of harm occurring.

Safety Freedom from unacceptable risk or hazard.

Safety circuit Any circuit or series of linked safety devices whose object is to provide protection from the hazards of a machine. Electrical safety circuits should be complete, i.e. made, before a machine can start and any break in it should cause the machine to come to a stop or revert to a safe condition.

Safety-critical Any guard, component, function or system whose continuing effectiveness is critical to the safe operation of the machine.

Safety-critical function The function of equipment or components whose continuing fault-free operation is essential for the continuing safe operation of a machine.

Safety device An equipment, device or means other than a guard that functions to prevent an operator from being harmed by a dangerous part of a machine.

Safety function That function of a component or equipment concerned with or contributing to a freedom from risk.

Safety integrity The ability of a component or system to fulfil its safety function and continue doing so.

Safety-related system Any guard, component or system which is required to perform a safety function aimed at reducing to an acceptable level any hazards or risks but is not in itself critical to providing that safety.

Safety valve The valve provided on all pressure vessels to ensure that the maximum operating pressure is not exceeded. Sometimes referred to as a *pressure relief valve*.

Safe working practices Methods of working, based on safe systems of work, that ensure any risk of injury or damage is eliminated or reduced to a minimum.

Scotch A mechanical strut or member inserted in a machine to physically prevent movement.

Slow break switch A switch in which the close and open action is directly controlled by the switch actuator.

Trip device A device which when actuated causes the machine to stop or revert to a safe condition. The device can be of mechanical, electrical or electronic operation.

Two-hand control An arrangement of two start buttons that requires instantaneous operation of both buttons before the machine will start. Release of either button causes the machine to stop and revert to a safe condition. Both buttons must be released before re-actuation of movement can occur. The two buttons to be located so they cannot be actuated by one hand or one hand and another part of the body.

Unexpected start-up Any start-up that is not expected whether initiated by an operator, as part of an automatic operation of the machine or other cause.

Appendix 3
Abbreviations

ac	Alternating current
Act	Part of a primary control element that actuates the switching
AOPDDR	Active opto-electronic devices responsive to diffuse reflections
bus	A system of electronic communication
CEN	Comité Européen de Normalisation
CENELEC	Comité Européen de Normalisation Electrotechnique
dc	Direct current
E/E/PE	Electrical, electronic and programmable electronic (controls)
EMC	Electro-magnetic compatibility
EMC-SLIM	Electro-magnetic compatibility – simpler legislation for the Internal Market
EN	European Normalisation (European standard)
ESPE	Electro sensitive protective equipment
EU	European Union
EUC	Equipment under control
FMEA	Failure mode and effect analysis
FSCS	Fault simulation of control systems
FTA	Fault tree analysis
HAVS	Hand-arm vibration syndrome
HAZAN	Hazard analysis
HAZOP	Hazard and operability studies
IEC	International Electrotechnical Commission (electrical standards)
ISO	International Standards Organisation (mechanical standards)
LPG	Liquified petroleum gas
LV	Low voltage
MOSAR	Method organized for a systemic analysis of risks
NDT	Non-destructive testing
Pce	Primary control element
PHA	Preliminary hazard analysis
PIPD	Passive infrared protective device

PLC	Programmable logic control
PSPE	Person sensing protective equipment
PSPD	Pressure sensing protective device
RA	Risk assessment
RCBO	Residual current circuit breaker with overcurrent protection
RCCB	Residual current circuit breaker
RCD	Residual current device
SIL	Safety integrity level
Sw	Switch, part of a primary control element that switches, i.e. causes a change of state
USB	Universal serial bus. A system that allows different electronic systems to communicate with each other
VDT	Visual display terminal
VDU	Visual display unit
WRULD	Work related upper limb disorder

Appendix 4

Smooth shaft pick up

No matter how smooth a rotating shaft may be, loose clothing and material will get caught and carried round by it. The reason for this is concerned with the dynamics of air movement.

Bernoulli's theorem states that a particle of water forming part of a larger body may be capable of doing work in virtue of its pressure, its velocity or its elevation. In considering the energy of any fluid system it is found that the sum of the three factors remains constant, i.e.

pressure energy + velocity energy + height energy = constant

If this is translated to a pneumatic system where the height energy is negligible since air has negligible weight, and expressed in terms of pressures the equation becomes:

$$P_{total} = P_{static} + P_{velocity}$$

Hence

$$P_{static} = P_{total} - P_{velocity}$$

Thus, where the air is not moving there is no velocity component of pressure and the static pressure equals the total pressure, whereas where the air is moving due to friction, i.e. adjacent to the shaft, there is a velocity component to its pressure.

This velocity component of pressure $P_{velocity} \sim \frac{1}{2}\rho v^2$ where ρ is the density of air and v is the air velocity.

Thus the faster the shaft rotates the faster the air adjacent to it moves and the greater is the dynamic component of the pressure. This creates a circumstance where the static air pressure at the shaft surface is less than that at a point, say, 50 mm away. This difference in pressure pushes the material on to the shaft and keeps it there as it wraps around the shaft.

This occurs no matter how smooth the shaft is.

Appendix 5

Pipeline colour codes

To prevent hazards arising from mistakes in, or lack of, knowledge of the contents of pipelines when carrying out work on them, the pipelines should be clearly identified. This can be by full colour along the whole length of the pipe or by bands or stripes of colour at regular intervals along its length and be supplemented by labels identifying the contents by name.

Colour coding for pipes:

Water	Green
Steam	Silver grey
Oil and combustible liquids	Brown
Gases – either as a gas or in liquid form	Yellow ochre
Acids	Violet
Air	Light blue
Other liquids	Black
Electrical services and ventilation ducts	Orange

Appendix 6

Permit-to-work

A permit-to-work should be used wherever the work to be undertaken presents a high risk of imminent injury or damage to the operator. The permit is aimed at ensuring that all possible precautions are taken before any work commences and requires discrete steps to be taken to accomplish this.

It is in five parts, each of which must be completed in sequence and at the appropriate stage of the work. Before commencement and at completion of each stage the permit must be countersigned by a responsible and authorized person.

If work on powered machinery when it is moving is involved, a person must be posted by the emergency stop controls or main isolator with instructions to isolate the machine in the event of an emergency.

The permit-to-work illustrated opposite is a simplified form but it contains the essential five parts. More complex forms can be developed from it to meet particular circumstances.

X Y Z Company Limited
PERMIT-TO-WORK

NOTES:
1. Parts 1, 2 and 3 of this Permit to be completed before any work covered by this permit commences and the other parts are to be completed in sequence as the work progresses.
2. Each part must be signed by an Authorized Person who accepts responsibility for ensuring that the work can be carried out safely.
3. None of the work covered by this Permit may be undertaken until written authority that it is safe to do so has been issued.
4. The plant/equipment covered by this Permit may not be returned to production until the Cancellation section (part 5) has been signed authorizing its release.

PART 1 DESCRIPTION
(a) Equipment or plant involved _____

(b) Location _____

(c) Details of work required _____

Signed _____   Date _____
person requesting work

PART 2 SAFETY MEASURES
I hereby declare that the following steps have been taken to render the above equipment/plant safe to work on: _____

Further, I recommend that as the work is carried out the following precautions are taken: _____

Signed _____   Date _____
being an authorized person

PART 3 RECEIPT
I hereby declare that I accept responsibility for carrying out the work on the equipment/plant described in this Permit-to-Work and will ensure that the operatives under my charge carry out only the work detailed.

Signed _____   Time _____   Date _____

Note: After signing it, this Permit-to-Work must be retained by the person in charge of the work until the work is either completed or suspended and the Clearance section (Part 4) signed.

PART 4 CLEARANCE
I hereby declare that the work for which this Permit was issued is now completed/suspended* and that all those under my charge have been withdrawn and warned that it is no longer safe to work on the equipment/plant and that all tools, gear, earthing connections are clear.

Signed _____   Time _____   Date _____
* delete word not applicable

PART 5 CANCELLATION
This Permit-to-Work is hereby cancelled

Signed _____   Time _____   Date _____
being a person authorized to cancel a Permit-to Work

Appendix 7

Protection of enclosures

Electrical equipment has to work in a range of different environments from which it must be protected if it is to give safe continuous service. The degree of protection needed will depend on the type and harshness of the environment in which it has to work. Different conditions demand different degrees of enclosure protection.

Enclosures are rated according to international standard IEC 60529, *Degrees of protection provided by enclosures,* and given an IP (International Protection) number. IP numbers are in two parts, the first concerned with protection against solid matter and bodies and the second with protection against liquids. The conditions for the various IP numbers are summarized below.

DEGREES OF PROTECTION OF ENCLOSURES

First digit	Degree of protection against foreign bodies	Second digit	Degree of protection against liquids
0	No protection	0	No protection
1	The ingress of large solid foreign bodies	1	The entry of droplets of condensed water
2	Medium sized solid foreign bodies	2	Drops of liquid falling at any angle up to 15° from the vertical
3	Small solid foreign bodies thicker than 2.5 mm	3	Rain falling at any angle up to 60° from the vertical
4	Small solid foreign bodies thicker than 1 mm	4	Liquids splashed from any direction
5	Dust in any amount sufficient to interfere with the operation of the equipment	5	Water spray or jet from any direction
6	Complete protection against the ingress of any dust	6	Against conditions on a ship's deck
–		7	Immersion in water
–		8	Indefinite immersion in water

Where no protection against the ingress of a liquid is required, the second digit 0, or sometimes X, is used, e.g. IP2X would provide protection against a medium sized foreign body, such as a finger, but not against a liquid.

The level of protection provided by the enclosure should be indicated on the equipment or a label permanently attached to it.

Some manufacturers use additional numbers to denote extra protection. Where this is done, the manufacturer should describe the level of additional protection provided.

Index

Abrasives, 18
Abrasive wheels, 24, 30
Abbreviations, 232
Access, 11, 215
Accumulators, hydraulic, 192
Air driers, 188
Air hose, 29
Air movers, 22
Air receiver, 29, 187, 188
Air system services, 185
Anthropometrics, 206
Anti static devices, 33
Arm reach, 111, 114

Bacon slicer, 29
Band saw, 26
Barrier:
   deterrent, 67
   physical, 107
Belt conveyor, 28, 68
Blades:
   doctor, 30
   cutting, 24, 30
Blood, 95, 102
Blowdown, 198
Body clearances, 116
Body movements, 96
Boilers, steam, 29, 193
Brakes, 68, 178
   electrodynamic, 120
      dc injection, 120
      regenerative, 120
      reverse plugging, 120
   mechanical, 119
   systems, 119

Burners, 196
Burns, 32
Burrs, 30

Cam operated interlock, 61, 71
Cams:
   linear, 62
   rotary, 61
Capacitive detection methods, 83
Carbon dioxide, 34
Car hoist, 181
Cartridge tools, 27
Chain and sprocket, 29
Chain saws, 27
Circuit fault protection, 129
Close proximity sensors, 83
Clutches, 121
   dog, 123
   hydraulic, 123
   magnetic, 123
   plate, 123
Colours:
   coding, 104
   fluorescent, 99
   pipeline codes, 235
   protocol, 98
   marking, 13
Combustors, 196
Common mode failures, 15
Compressed air, 185
Compressor, 186
Controls, 96
   pedal, 97
Counter balance weights, 25, 29, 118
Counter rotating rolls, 21

Cranes:
    mobile, 179
    overhead electric traveling, 179
    tower, 178
Crawl control, 80
Cross monitoring, 77, 157
Cutters:
    milling, 29
    rotary, 30
Cutting blades, 24
Cylinders, pneumatic, 29

Decommissioning, 38
Degree of exposure, 46
Delay start, 93
Delphi Technique, 38
Design hazard reduction, 36
Design rationale, 17
Design risk assessment, 38
Detection:
    capablility, 81
    zone of, 81
Diode link, 76
Diversity, 15, 71, 152
Draw rolls, 21
Dual channel interlocking, 76
Duplication, 15
Dust, 30, 31, 215

Earth, 31, 132
Earth fault protection, 131
Earth leakage protection, 133
Ejection of material, 13, 18, 30
Electrical, electronic and programmable electronic systems (E/E/PE), 49
Electricity, 19, 31
Electrosensitive protective equipment, 62, 81, 87
Electric shock, 202
Emissions, 200
Energy:
    stored, 19
    contained, 29
    potential, 29
Environment, 12, 210
Ergonomic principles, 11
Ergonomics, 95

Emergency stop:
    circuits, 149
    switch, 69, 80, 89, 98, 181
Escalators, 181
Expanded metal, 108
Eye protection, 30

Fans:
    axial flow, 22
    radial flow, 22
Failure Mode and Effect Analysis (FMEA), 38
Failure rate frequency, 41
Fatigue, 100
Fault Simulation of Control Systems (FSCS), 38
Fault Tree Analysis (FTA), 38
Feed stations, 117
Feed water, 197
Fence, 65
Fencing, distance, 113
Fibre optics, 87
Fieldbus systems, 173
Figures, 103
Fire:
    fixed fighting system, 181
    hazard, 30
    precautions, 30
Fixed structures, 25
Flicker, 33
Floors, 216
Foundations, 216
Fly press, 27
Foot reach, 112
Fork lift trucks, 181
Fuel, 195
Fumes, 31, 215

Gamma rays, 102
Gas burning, 30
Gauges, 97
Gauging, 126
Gears, meshing, 23
Glare, 33
Grab wire, 68, 88
Guard, 56, 68, 107
    adjustable, 60
    automatic, 63

Guard – *cont.*
  control, 62
  delayed opening, 145
  distance, 66
  enclosing, 107
  fixed, 57
  gaps in, 110
  interlocking, 61
  irregular openings, 113
  locking, 75
  material, 108
  movable, 57, 71, 97
  openings, 110, 206
  push away, 63
  removable, 59, 96, 97
  sliding, 61, 127
  self adjusting, 62
  tunnel, 63, 117
Guillotine blades, 29

Hand-arm vibration syndrome (HAVS), 32
Hand reach, 111
Hazard, 18, 35
Hazard and Operability Studies (HAZOP), 38
Hazard analysis (HAZAN), 38
Hazard elimination, 39
Hazard identification, 38
Hazard reduction, 39
Hazard zone, 71
Hazardous chemicals, 31
Hedge trimmers, 30
Hoist, car, 181
Hold-to-run control, 68, 80, 149
Humidity, 101
Hydraulic fluid, 191
Hygiene, 12, 216

Illuminance, 99
Illumination, levels of, 99
Indicating instruments:
  rotary, 103
  linear, 103
  digital, 103
Indicators:
  lights, 103, 104
  performance, 103
  qualitative, 102
  quantitative, 102
Induction heating, 33
Inductive detection methods, 83
Inertia, 68
Information, 40
Infrared sensing, 83
Instruction manuals, 13
Instruments, 97, 102, 103
Insulation failure, 130
Integrity, 16
Integrity of interlocking guards, 12
Interlocking:
  control, 53, 76, 77, 78, 135, 154
  devices, 71, 76
  guards, 68
  power, 53, 76, 77, 78, 136, 157
  systems, 76
Interlocking media:
  electrical, 77
  hydraulic, 78, 154
  mechanical, 77
  pneumatic, 79
Interlocking switches:
  captive key, 94
  electronic, resonance coded, 78
  key exchange, 146
  linear cam operated, 61, 71, 78
  rotary cam operated, 61, 71, 78
  negative mode, 73
  opposed mode, 76
  positive mode, 71, 73, 76
  tongue operated, 74, 78
Internal combustion engines, 34

Jigs and fixtures, 70, 207

Key exchange system, 68, 91, 146
Knives, rotary, 24

Labels, 207
Lamps:
  fluorescent, 34, 99
  incandescent, 34
Lasers, 33, 68, 70, 83, 85, 102

Latent heat of steam, 31
Leg reach, 112
Legislative requirements, 11
Letters, 103
Levels of risk, 49
Lifting, 183
 accessories, 177, 182
 equipment, 177
  handling, 183
  stability, 183
  transporting, 183
 machinery, 177
Lifts, 27, 180
 paternoster, 181
 radio control, 182
 remote control, 182
 scissor, 180
Lighting, 12, 33, 99, 212
Lighting intensity, 99
Limited movement control, 80
Limiting devices, 94
Linear sliding movement, 18
Linisher, 25
Locking off, 203
Lock out condition, 83
Loose sleeve, 20
Low voltage, 202
Lubrication, 215
Luminaires, 99
Lux, 99

Machine specification, 11
Machines, hand fed, 96
Maintainability, 15
Maintenance, 16, 100, 214
Maintenance factor, 46
Manuals:
 instruction, 16, 40
 maintenance, 40
Material handling, 214
Method Organized for a Systemic Analysis of Risks (MOSAR), 38
Microwave, 33
Monitor, 40
Monitoring, safety circuit, 137
Mowing machine, 30
Multi-operator machines, 13
Muting, 81

Nail guns, 27
Negative mode operation, 73
Noise, 32, 100, 213
 reduction, 100
Notices, 216

Operate with guards open, 12
Operating manuals, 16, 206
Operational factor, 46
Operator, 10
Over current protection, 91

Pearce method for determining SIL, 41, 46
Perforated metal, 108
Periodic examination, 184
Permit-to-work, 202, 236
Person sensing devices, 80
Personal protective equipment (PPE), 13, 70
Personnel sensing protective equipment, 81
Perspex, 108
Photo-electric curtain, 62, 68
Photo-electric sensing, 83
Physical agents, 19
Physiology, 95
Pipes:
 flexible, 189
 leaking, 30
Pipework, 188, 192, 198
Plastic, thermosetting, 110
Pneumatic:
 installations, 162
 safety circuits, 164
 valves, 163
Polycarbonate, 108
Portable (electrical) equipment, 203
Positive break switch, 74
Positive mode operation, 71, 73
Power clamp, 118
Power take off (PTO), 30
Preliminary Hazard Analysis (PHA), 38
Pressure, 102
 intensification, 161, 190
 maximum operating, 190
 surge, 190

Pressure reducing valves, 162, 199
Pressure relief valves, 184, 188, 193, 195
Pressure sensitive edges, 88
Pressure sensitive wires, 88
Pressure sensitive mats, 83, 86
Probability, 46
Programmable logic controllers, 173
Projections, 21
Prop, 67
Protection of enclosures, 238
Pulley belts, 28
Pumps, hydraulic, 191

Quality assurance, 13
Quality assurance system, 14
Quantitative risk assessment, 48

Rack and pinion gears, 27
Radar sensing, 83
Radiations, 19, 102
   infrared, 33, 102
   ionizing, 32, 102
   microwave, 102
   non-ionizing, 32
   radio frequency, 33, 102
   ultraviolet, 33, 102
   X-ray, 33
Reach, safety distances, 115
Relay logic, 247
Reliability, 14, 16
Redundancy, 15, 71, 152
Repetitive actions, 32, 101
Ridley method for determining SIL, 41, 47
Residual current device (RCD), 30
Risk, 35
   levels of, 76
Risk assessment, 35, 39, 201
Risk rating, 48
Risk reduction, 12, 35, 40
Risk reduction strategy, 36
Rivetting machines, 26
Robots, 29, 68
Roller conveyors, 22
Rolling wheels, 28
Rotation, 18, 20
Routers, 24

Safe by position, 11
Safe operating procedures, 12
Safeguarding device, 56
Safeguarding methods, 12
Safe position, 71
Safe working load, 178
Safe working pressure, 184
Safe systems of work, 13, 70
Safety circuit:
   electronic, 146
   hydraulic, 153
   monitoring, 137
   dual channel, 139, 157, 169
   multi channel, 149
   single channel, 135, 154, 166
Safety clothing, 208
Safety control circuits, 133
Safety critical, 15, 31, 126, 166
Safety culture, 11
Safety distances, 83, 115
Safety gaps, 115
Safeguarding, 11
Safeguarding system selection, 49
Safety background, 11
Safety catches, 118
Safety clothing, 208
Safety Integrity Level (SIL), 40 et seq.
Safety limit trip, 179
Safety policy, 11
Safety related, 15, 31, 126, 166
Safety systems:
   electrical, electronic and programmable electronic systems, 49
   electro-mechanical systems, 53
   hydraulic and pneumatic systems, 53
Saws:
   circular, 29
   band, 26
Scanning cycle, 83
Scissor lifts, 26, 180
Scotch, 66
Sensor, 81
Services, 211
Severity, 46
Servicing, 16
Shadows, 33
Shafts, rotating, 68, 234
Sharp edges, 19, 29

Shock, 31
Signs and signals, 208, 216
Single channel interlocking, 76, 166
Smoke alarms, 33
Sound proof enclosure, 186
Space, 210
Speed, 102
   of approach, 83
Spillages, 13
Spoked wheels, 23
Start-up delay devices, 12
Steam generators, 193, 197
Stopping performance monitoring, 81, 139
Stored energy, 175, 208
Stress, 100
Stroboscopic effect, 33
Supervision, 13, 207
Switch:
   fast break, 126
   foot operated, 126
   interlocking, 127
   key interlock, 93
   limit, 76, 126
   magnetic, 76
   micro, 126
   positive break, 74, 76
   proximity, 90, 126
   start, 97
   telescopic trip, 90
Switching contacts, 128
Systems of work, 201

Take-off stations, 117
Technical file, 4
Temperature, 31, 101, 102, 212
Terms, glossary, 227
Test instruments, 203
Time delay guard locking, 68, 75, 93
Total enclosure, 68
Tower cranes, 178

Training, 13, 16
Trip devices, 69
Two-hand control, 34, 79

Ultra sonic sensing, 83

Vacuum pumps, 187
Validation, 17
Valves:
   control, 78
   interposed, 78
Vee belt drives, 108
Ventilation, 101, 211
Ventricular fibrillation, 31
Vibrations, 32, 100, 214
Vibration white finger, 32, 100
Vision, 101

Warning:
   audible, 126
   notices, 31, 101
Waste, 214
Waste removal, 13
Water jetting, 34, 70
Welding, 30
Weld mesh, 108
'WHAT IF' technique, 38
Wire stitching, 26
Work holding devices, 118
Work related upper limb disorder (WRULD), 32, 101
Working load limit, 178
Working, rates of, 100

X-rays, 102

Zone of detection, 81, 83